Collins

Cambridge Lower Secondary

Science

**PROGRESS BOOK 8:
TEACHER PACK**

Series editor: David Martindill
Authors: Aidan Gill, Emma Poole and Heidi Foxford

William Collins' dream of knowledge for all began with the publication of his first book in 1819.

A self-educated mill worker, he not only enriched millions of lives, but also founded a flourishing publishing house. Today, staying true to this spirit, Collins books are packed with inspiration, innovation and practical expertise. They place you at the centre of a world of possibility and give you exactly what you need to explore it.

Collins. Freedom to teach.

Published by Collins

An imprint of HarperCollins*Publishers*
The News Building, 1 London Bridge Street, London, SE1 9GF, UK

HarperCollins*Publishers*
Macken House, 39/40 Mayor Street Upper, Dublin 1, D01 C9W8, Ireland

Browse the complete Collins catalogue at
collins.co.uk

© HarperCollins*Publishers* Limited 2024

10 9 8 7 6 5 4 3 2 1

ISBN 978-0-00-867936-1

All rights reserved. No part of this publication may be reproduced, stored in a retrieval system, or transmitted in any form by any means, electronic, mechanical, photocopying, recording or otherwise, without the prior written permission of the Publisher or a licence permitting restricted copying in the United Kingdom issued by the Copyright Licensing Agency Ltd, 5th Floor, Shackleton House, 4 Battle Bridge Lane, London SE1 2HX.

British Library Cataloguing-in-Publication Data

A catalogue record for this publication is available from the British Library.

The questions, accompanying marks and mark schemes included in this resource have been written by the author and are for guidance only. They do not replicate examination papers and the questions in this resource will not appear in your exams. In examinations the way marks are awarded may be different. Any references to assessment and/or assessment preparation are the author's interpretation of the syllabus requirements.

This text has not been through the endorsement process for the Cambridge Pathway. Any references or materials related to answers, grades, papers or examinations are based on the opinion of the author. The Cambridge International Education syllabus or curriculum framework associated assessment guidance material and specimen papers should always be referred to for definitive guidance.

Series Editor: David Martindill
Authors: Aidan Gill, Emma Poole and Heidi Foxford
Publisher: Elaine Higgleton
Product manager: Catherine Martin
Product developer: Roisin Leahy
External Project Manager: Just Content Ltd
Development editor: Rebecca Ramsden
Copyeditor: Nick Hamar
Proofreader: Just Content Ltd
Cover designer: Gordon MacGilp
Cover illustrator: Ann Paganuzzi
Typesetter: PDQ Digital Media Solutions Ltd
Production controller: Sarah Hovell and Lyndsey Rogers
Printed and bound by Ashford Colour Press Ltd

We are grateful to the following teachers for providing feedback on the resources as they were developed: Dr. Rahul Sharma at IRA Global School, Mumbai, Mr Frank Akrofi and Mr Samuel Yeboah, Dániel Szücs at International School of Budapest, Ms Shalini Reddy at Manthan International School and Ms Sejal Vasrkar at SVKM JV Parekh International.

This book contains FSC™ certified paper and other controlled sources to ensure responsible forest management.

For more information visit: www.harpercollins.co.uk/green

Contents

Introduction	v
Respiration and movement	1
Self-assessment and reflective learning page	6
Nutrition	8
Self-assessment and reflective learning page	13
Ecosystems	15
Self-assessment and reflective learning page	20
Materials and solutions	22
Self-assessment and reflective learning page	28
Chemical changes	30
Self-assessment and reflective learning page	36
Motion	38
Self-assessment and reflective learning page	43
Forces	45
Self-assessment and reflective learning page	52
Light and magnets	54
Self-assessment and reflective learning page	60
The Earth in space	62
Self-assessment and reflective learning page	67
End of Year Test 1	69
End of Year Test 2	81

Mark scheme Respiration and movement	93
Mark scheme Nutrition	95
Mark scheme Ecosystems	97
Mark scheme Materials and solutions	99
Mark scheme Chemical changes	101
Mark scheme Motion	103
Mark scheme Forces	105
Mark scheme Light and magnets	107
Mark scheme The Earth in space	109
Mark scheme End of Year Test 1	111
Mark scheme End of Year Test 2	115
Periodic Table	120
Glossary	121

Introduction

This *Stage 8 Progress Teacher Pack* (and the *Stage 8 Progress Student's Book*) can be used to support the *Collins Cambridge Stage 8 Lower Secondary Science course* or to supplement your own resources. The *Progress Teacher Pack* contains

- nine End of Unit Tests offering practice questions to assess understanding of the Lower Secondary Science course
- two summative End of Year Tests
- student Self-assessment sheets for each of the End of Unit Tests
- mark schemes for each of the End of Unit Tests and for the End of Year Tests.

How to use the Progress resources

This downloadable, editable and photocopiable Teacher Pack contains a range of End of Unit Tests that are designed to be valuable and flexible formative and summative assessment resources. They can be used to identify strengths and weaknesses and to pinpoint how future teaching should be adjusted to ensure all students make good progress.

The nine End of Unit Tests can be used as class tests or can be set for students to complete at home. They can be set at the end of a unit of teaching or can be combined to create a longer end of term test if appropriate.

Some of the questions in each End of Unit Test are written to address the Cambridge Thinking and Working Scientifically Learning Objectives:

- Models and representations
- Scientific enquiry: purpose and planning
- Carrying out scientific enquiry
- Scientific enquiry: analysis, evaluation and conclusions.

Each End of Unit Test is designed to be marked out of 20. Teachers may wish to set a time limit of 20 minutes.

The End of Year Tests assess objectives taught across the whole year. The style of the End of Year Tests is otherwise the same as the End of Unit Tests, with a mixture of question styles and question difficulties, as well as the inclusion of some Thinking and Working Scientifically questions. Questions are also set in the context of practicals where appropriate, ensuring that students have experience of answering questions on investigative work. These tests could be used for summative purposes as end of year examinations or as practice for students ahead of their examinations. The End of Unit Tests are separated out into the different science subjects, but the End of Year Tests cover a combination of the different science subjects.

The *Progress Teacher Pack* includes clear mark schemes for each End of Unit Test and for the End of Year Tests. These mark schemes contain notes on what should be seen for full marks to be awarded. They also set out how part marks can be awarded in a question where the full correct answer is not reached.

The student Self-assessment sheets give students the opportunity to reflect on their understanding. Students are given a list of statements about their understanding of the course content and are able to rank their understanding of the statement between 'I don't know', 'I need more practice' and 'I understand.' This will provide both students and teachers with a relatively quick way to assess the student's overall confidence with the content. They are also invited to rank their understanding of the Thinking and Working Scientifically content. Students are then invited to produce a written reflection to answer these questions: 'What went well in this topic?', 'What could you do better next time?' and 'What parts of the course could your teacher go through in a revision lesson which would support your improvement?'. There is also a space for teachers to make a comment about the student's understanding of the content.

Teachers can use the results of the End of Unit Tests and the students' Self-assessment sheets to help them in future lesson planning. For example, if many students struggled with work linked to the internal structure of the Earth, a teacher may wish to bear this in mind when planning their teaching of a related topic, such as seismic waves – teachers could, for instance, include starter activities recapping the earlier work.

Key features: End of Unit and End of Year Tests

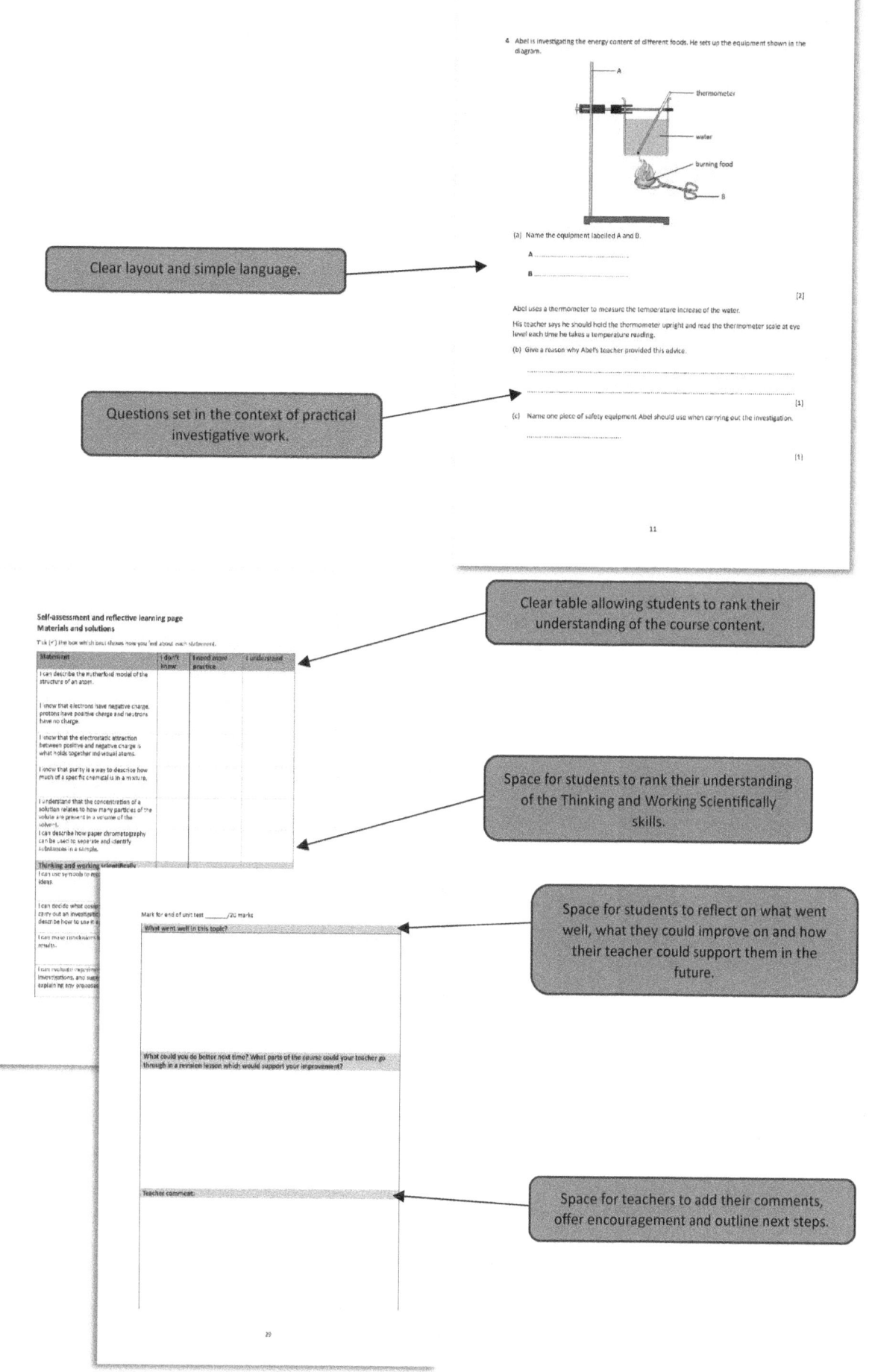

Key features: Mark schemes

Clear indication of the correct answer and number of marks to be awarded.

Information about which spec point the question relates to including Thinking and Working Scientifically statements.

Questions also linked to Assessment Objectives so that teachers can track which skills students need more support with.

Guidance given about how to award part marks

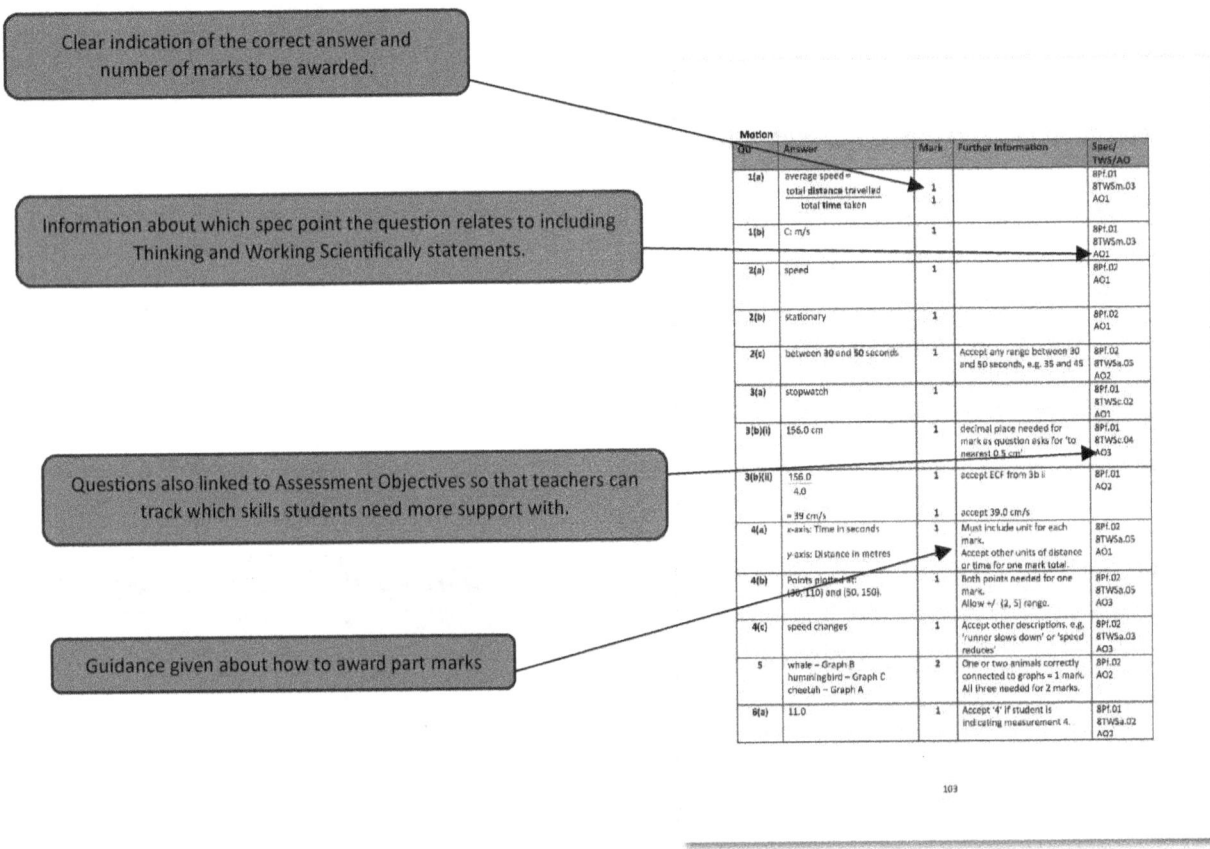

# End of Unit Test: Respiration and movement
Total = 20 marks

Name: .... Class: ....
Date: ....

1   The diagram shows the bones and muscles in the upper leg.

(a)   The joint in the knee allows forwards and backwards movement of the lower leg.

Circle the name of this type of joint.

hinge   ball and socket   pivot

[1]

(b)   The quadriceps and hamstring muscles are antagonistic.

Complete the table by writing **contracts** or **relaxes** to show the action of the quadriceps and hamstring muscles in each type of movement.

Muscle	Bending the leg	Straightening the leg
quadriceps		
hamstrings		

[2]

2 Draw a line to match each **component of blood** to its **function**.

Component of blood   **Function**

plasma — transports oxygen

red blood cell — protects against pathogens and fights disease

white blood cell — transport cells, nutrients and carbon dioxide

[2]

3 The diagram shows an alveolus in the lung.

(a) Give the name of **gas A**. ....

[1]

(b) Name the process by which gas moves from the alveolus into the blood capillary.

....

[1]

(c) Explain **one** way the structure of the alveolus blood capillary is related to its function.

....

[1]

Inhalation requires movement of several structures within the respiratory system.

(d) Explain what must happen to the diaphragm for air to move into the lungs.

....

....

[2]

4 Respiration takes place in all living organisms.

(a) Where does aerobic respiration occur? Tick (✓) **one** box.

nuclei of muscle cells ☐

blood ☐

mitochondria ☐

alveoli ☐

[1]

(b) Complete the summary word equation for aerobic respiration.

oxygen + ... ⟶ carbon dioxide + water

[1]

(c) Long-distance runners can improve their performance by drinking sports drinks that contain a type of sugar that can be quickly absorbed into the blood.

Suggest how drinking this type of sports drink could improve performance in a long-distance running race.

....

....

....

[2]

5 Annika investigates the effect of exercise on heart rate.

She records her heart rate while skipping for 60 seconds.

A heart rate monitor is used to measure and record her heart rate as she skips.

The heart rate monitor uses a chest strap which detects the electrical signals of the heart as it beats and records it every 10 seconds.

The graph shows Annika's results.

(a) Write down Annika's heart rate when she stops skipping at the end of the minute.

....beats per minute

[1]

(b) Explain the pattern shown by the graph between 10 and 60 seconds.

....

....

....

....

[3]

(c) It is possible to measure the heart rate by taking the pulse in the neck or the wrist.

(i) Explain why using a heart rate monitor is a better method of measuring the heart rate.

....

....
[1]

(ii) Write down what Annika should do to make her results more reliable.

....

....
[1]

Self-assessment and reflective learning page
Respiration and movement

Tick (✓) the box which best shows how you feel about each statement.

Statement	I don't know	I need more practice	I understand
I can identify ball-and-socket and hinge joints in diagrams.			
I can explain how antagonistic muscles work together to move the bones of a hinge joint.			
I know that red blood cells carry oxygen, white blood cells help protect against pathogens and plasma transports blood cells, nutrients, and carbon dioxide.			
I can describe how the structure of the human respiratory system is related to its function of gas exchange.			
I can describe the diffusion of oxygen and carbon dioxide between blood and the air in the lungs.			
I know that aerobic respiration takes place in the mitochondria inside cells.			
I know the word equation for aerobic respiration.			
Thinking and working scientifically			
I can evaluate whether measurements and observations have been repeated sufficiently to be reliable.			
I can explain why it is important to take measurements that are accurate and precise.			
I can describe trends and patterns in results.			
I can make conclusions by interpreting results.			
I can evaluate experiments and investigations, and suggest improvements, explaining any proposed changes.			

Mark for end of unit test _____ /20 marks

What went well in this topic?

What could you do better next time? What parts of the course could your teacher go through in a revision lesson which would support your improvement?

Teacher comment:

End of Unit Test: Nutrition
Total = 20 marks

Name: .... Class: ....

Date: ....

1 This question is about a balanced diet.

   (a) Complete the sentence.

   A balanced diet contains proteins, ..., fats and oils, water,

   minerals and ....

   [2]

   (b) Circle the food that is a good source of protein.

   chicken   **orange**   **cauliflower**   **water**   **olives**

   [1]

   (c) Which sentence best describes a balanced diet? Tick (✓) **one** box.

   A diet containing no fat or sugar. ☐

   A diet that is very high in fibre. ☐

   A diet that contains all the nutrients in the recommended proportions. ☐

   A diet that has a very high percentage of fruit and vegetables. ☐

   [1]

   (d) Draw a line from each **food group** to its matching **function**.

Food group	Function
fats and oils	to provide insulation
proteins	to provide energy
carbohydrates	for growth and repair

   [2]

2 The chart shows the vitamin C content of different types of food.

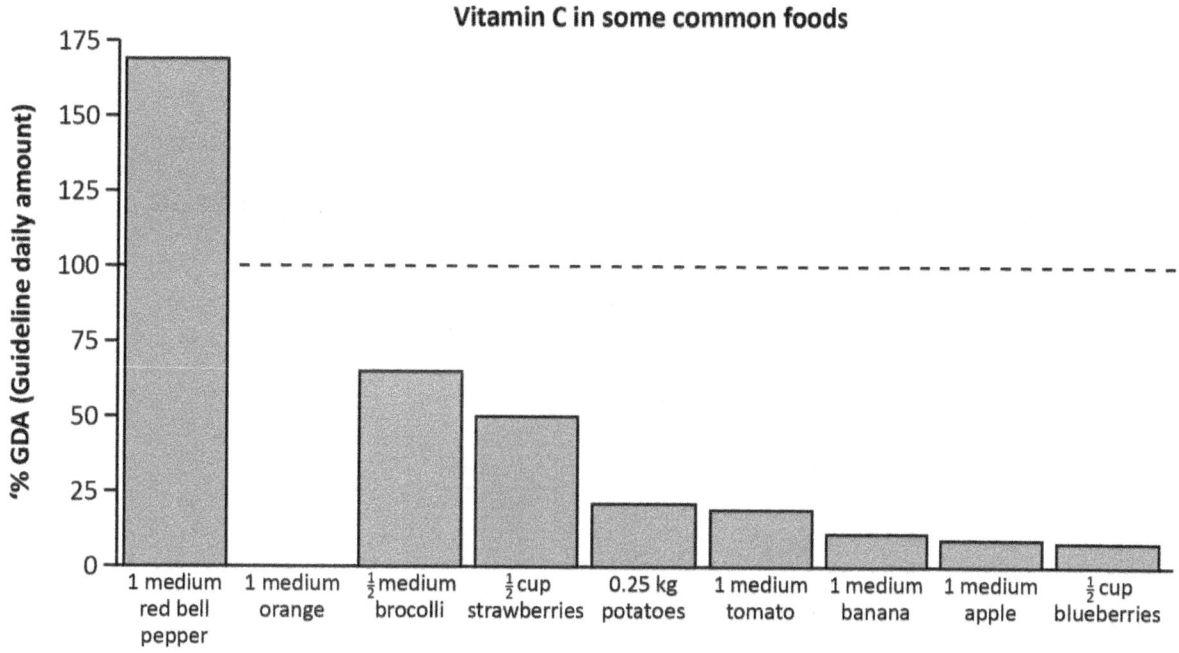

One medium orange contains 75% of the recommended daily amount for an adult.

(a) Draw a bar on the chart to show the percentage of vitamin C from one medium orange.

[2]

Emilia starts her day by eating half a cup of strawberries.

(b) Use the graph to find out what percentage of the recommended daily amount Emilia has eaten.

....%

[1]

(c) The recommended daily amount of vitamin C is 40 mg (milligrams).

A kumquat is a type of citrus fruit native to South China. On average one kumquat contains 8 mg of vitamin C.

Calculate how many kumquats a person would need to eat to reach the recommend daily amount of vitamin C.

....

[1]

(d) Describe the importance of including vitamin C in a balanced diet.

....
[1]

3 The image shows a nutritional label for an orange flavoured drink made for children.

The company that produces the drink advertise it as being a good source of vitamin C and minerals.

Nutrition Facts	
6 servings per container	
Serving size 1 packet (2g)	
Amount per serving	
	GDA*
Total Fat 0g	0%
Total Sugars 39g	46%
Protein 0g	
Calcium 60mg	4%
Vitamin C	3%
Not a significant source of saturated fat, trans fat, cholesterol, dietary fibre, vitamin D, iron and pottassium.	
*GDA = Guided daily amount	

Mia says the drink is a healthy choice because it contains lots of vitamins and minerals.

Imran says the drink is not a healthy choice.

Do you agree with Mia or Imran? Use evidence from the label to support your answer.

....

....

....
[2]

4 Abel is investigating the energy content of different foods. He sets up the equipment shown in the diagram.

(a) Name the equipment labelled A and B.

A ....

B ....

[2]

Abel uses a thermometer to measure the temperature increase of the water.

His teacher says he should hold the thermometer upright and read the thermometer scale at eye level each time he takes a temperature reading.

(b) Give a reason why Abel's teacher provided this advice.

....

....

[1]

(c) Name one piece of safety equipment Abel should use when carrying out the investigation.

....

[1]

(d) Suggest a piece of equipment he could use to measure the temperature that would make his results more accurate.

....

[1]

5 The table shows data on the percentage of males that smoked between 1980 and 2005 in the UK.

Year	Percentage of males who smoke (%)
1980	55
1985	46
1990	39
1995	37
2000	32
2005	28

(a) Describe the trend shown by the data in the table.

....

....

[1]

Smoking is harmful to health. Many people die from smoking-related illnesses each year.

(b) Name **one** disease linked to smoking.

....

[1]

Self-assessment and reflective learning page
Nutrition

Tick (✓) the box which best shows how you feel about each statement.

Statement	I don't know	I need more practice	I understand
I know that a balanced diet for humans includes carbohydrates, protein, fats and oils, water, minerals and vitamins.			
I understand that carbohydrates and fats can be used as a store of energy in animals.			
I know the function of proteins, carbohydrates, fats and oils, and vitamin C in the diet.			
I understand how human health can be affected by smoking.			
Thinking and working scientifically			
I can decide what equipment is required to carry out an investigation or experiment and describe how to use it appropriately.			
I can describe how to take appropriately accurate and precise measurements.			
I can evaluate a range of secondary information sources for their relevance.			
I can describe trends and patterns in results.			
I can evaluate experiments and investigations and suggest improvements.			
I can present and interpret observations and measurements appropriately.			

Mark for end of unit test _____/20 marks

What went well in this topic?

What could you do better next time? What parts of the course could your teacher go through in a revision lesson which would support your improvement?

Teacher comment:

End of Unit Test: Ecosystems
Total = 20 marks

Name: .... Class: ....

Date: ....

1 (a) Complete the sentences to describe the difference between an ecosystem and a habitat.

A habitat is where a type of organism ....

An ecosystem is all the organisms and all the .... in a certain type of habitat.

[2]

(b) Look at the list of different types of ecosystems.

Use words from the list to identify each type of ecosystem shown in the diagrams below.

desert

polar

rainforest

savannah

wetland

....

[2]

2 Pritesh investigates the choice of habitat by woodlice. He wants to find out which type of habitat woodlice prefer.

Pritesh collects the woodlice from the school garden

(a) Circle the name of the equipment best suited to collecting woodlice.

pond net   pooter   quadrat   sweep net   trowel

[1]

A choice chamber is a piece of equipment used to investigate how small organisms respond to different environmental conditions. The choice chamber has four sections each with different conditions. The woodlice can move freely between each section.

(b) Name the independent variable in Pritesh's investigation.

....

[1]

Pritesh uses a drying chemical called anhydrous calcium chloride to remove the moisture from two chambers. Anhydrous calcium chloride is a skin irritant.

(c) Describe **one** way Pritesh could minimise risk of skin irritation when handing the anhydrous calcium chloride.

....

[1]

Pritesh puts 20 woodlice into the choice chamber and uses a stopwatch to time 5 minutes.

The diagram shows the position of the woodlice after the end of the first 5 minutes (Trial 1).

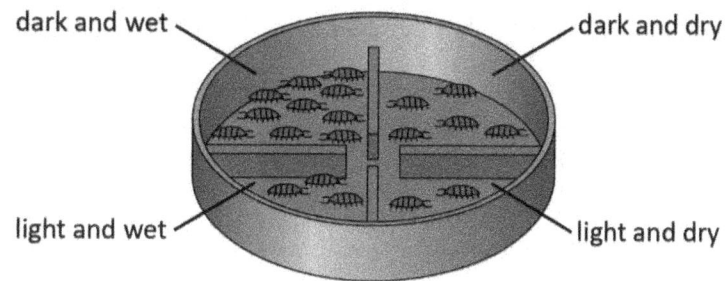

(d) Complete the table to record the results of Pritesh's investigation.

Trial				
1				
2				

    (i)   Write an overall heading and four subheadings for each of the five empty columns shaded grey.

[2]

    (ii)   Write the results from the first trial by using the information in the diagram.

[1]

(e) Identify the correct conclusion from this investigation.

The woodlice prefer the dry and light conditions. ☐

The woodlice do not show any preference between the different conditions. ☐

The woodlice prefer the wet and dark conditions. ☐

The woodlice prefer the dry and dark conditions. ☐

[1]

3 Cane toads are native to Central and South America.

In the 1930s, Cane toads were introduced to Australia to control pests that ate sugar cane.

Cane toads are poisonous to animals that eat them.

They are predators and their introduction resulted in a decrease in the population of native species such as the yellow spotted monitor lizard.

Cane toads are now considered an invasive species in Australia.

(a) Describe what is meant by the term 'invasive species.'

....

.... [1]

(b) Explain why the cane toad is considered to be an invasive species and not just non-native.

....

.... [1]

The diagram shows a cane toad food web.

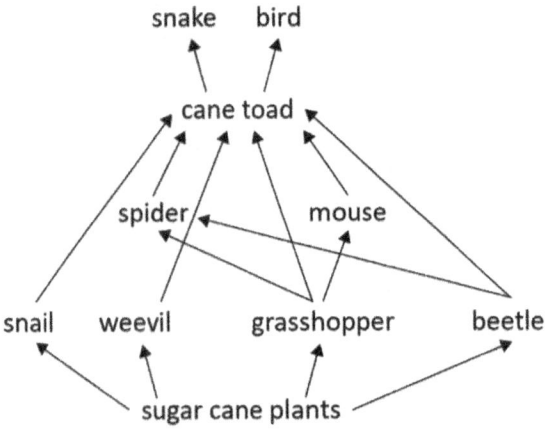

(c) Describe what the arrows represent in the food web.

....
[1]

(d) Give **two** ways an increase in the population of cane toads could decrease the population of spiders.

1..

2..
[2]

4 Jasminder is studying bioaccumulation of mercury in an ocean food chain.

krill ⟶ oyster ⟶ tuna ⟶ shark

In this food chain name:

(a) a secondary consumer ....

[1]

(b) the top predator. ....

[1]

Jasminder investigated the concentration of mercury in the tissues of each organism in the food chain.

Her results are shown in the table.

Organism	Concentration of mercury (arbitrary units)
krill	1
oyster	3
tuna	9
shark	28

(c) Explain why the shark has the highest concentration of mercury.

....

....

....

[2]

Self-assessment and reflective learning page
Ecosystems

Tick (✓) the box which best shows how you feel about each statement.

Statement	I don't know	I need more practice	I understand
I can identify different types of ecosystems, for example, rainforest and desert.			
I can describe the impact of the bioaccumulation of toxic substances on an ecosystem.			
I can describe how a new and/or invasive species can affect other organisms and an ecosystem.			
Thinking and working scientifically			
I can plan a range of investigations of different types, while considering variables appropriately.			
I can make risk assessments for practical work to identify and control risks.			
I can describe the equipment required to carry out an investigation.			
I can describe how to carry out practical work safely.			
I can record measurements in an appropriate form.			

Mark for end of unit test _____ /20 marks

What went well in this topic?

What could you do better next time? What parts of the course could your teacher go through in a revision lesson which would support your improvement?

Teacher comment:

End of Unit Test: Materials and solutions
Total = 20 marks

Name: .... Class: ....

Date: ....

1   This question is about pure substances.

Use words in the box to complete the sentences below.

| melting range | melting point | mixture |
| element | alloy | compound |

A pure substance contains only one type of ....or one type of

....

Pure substances have a precise ....

[2]

2   Atoms contain three types of sub-atomic particle: protons, neutrons and electrons.

(a)   Draw a line to link each **sub-atomic particle** to its correct **charge**.

Sub-atomic particle   **Charge**

proton   no charge

neutron   positive

electron   negative

[2]

Atoms are held together by strong forces of attraction between positively and negatively charged particles.

(b)   Circle the type of attraction that occurs inside atoms.

   magnetic   electrostatic   ionic   molecular

[1]

3 Hydrogen peroxide is a useful chemical. Dilute solutions of hydrogen peroxide can be used to bleach hair.

The symbol

○

is used to represent hydrogen peroxide molecules.

The symbol

●

is used to represent water molecules.

(a) Write down the letter, **A**, **B**, **C** or **D**, that shows the highest concentration of hydrogen peroxide molecules.

....
[1]

(b) Which one of these units could be used to measure the concentration of hydrogen peroxide in a solution?

Tick (✓) **one** box.

grams ☐

g/cm^3 ☐

Joules ☐

Newtons ☐

[1]

4 Chromatography can be used to separate mixtures of coloured substances.

(a) Place these steps in order by writing **1** to **7** in the table to show how to carry out a chromatography experiment. The first three steps have been done for you.

Draw a pencil line 0.5 cm from the bottom of the chromatography paper.	1
Mark where the solvent has reached.	6
Add a small amount of solvent to the beaker.	3
Place the chromatography paper to the beaker.	4
Place a spot of the sample on the pencil line.	2
Allow the paper to dry.	7
When the solvent reaches near to the top of the paper remove the paper.	5

[2]

(b) Explain why the solvent front is marked onto the chromatogram when it is removed from the beaker.

....

....

[1]

(c) Explain why it is important that a pencil is used to draw a line on the chromatography paper.

....

....

[1]

(d) Douglas carries out an experiment to find out which colours are found in food colouring **X**.

Look at the results of his experiment shown below.

(i) Identify the name of the most soluble ink.

....
[1]

(ii) Identify the colours that have been mixed together to make food colouring **X**.

....

....
[1]

(iii) Suggest why the blue ink did not move.

....

....
[1]

(iv) Suggest what Douglas could do to see if the blue ink is a mixture.

....

....
[1]

(e) In a different chromatography experiment, Sabrina uses the R_f value to identify the inks present in a sample.

She calculates the R_f value using the equation:

$$R_f = \frac{\text{distance the ink moves}}{\text{distance the solvent moves}}$$

Her results are shown in the table below.

Distance the solvent moves = 12.0 cm
Distance the yellow ink moves = 3.5 cm
Distance the pink ink moves = 10.0 cm
Distance the purple ink moves = 3.0 cm
Distance the brown ink moves = 7.0 cm
Temperature = 25 °C
Width of the paper = 5.0 cm

Calculate the R_f value for the purple ink.

....

....

[2]

5 Our ideas about atoms have changed over time.

(a) The diagram below shows four models of an atom. Identify the diagram (**A**, **B**, **C**, or **D**) that shows the earliest model of the atom.

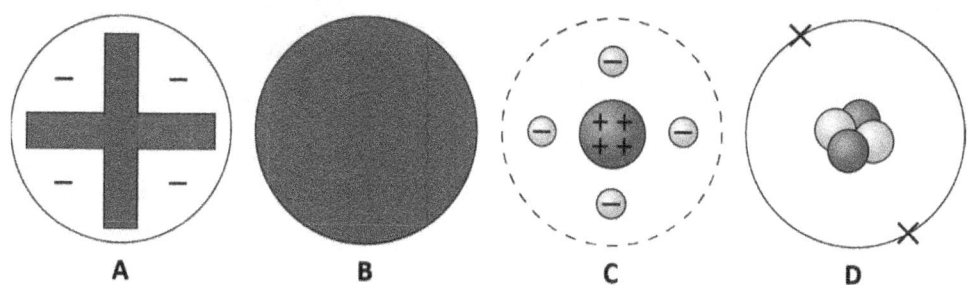

....
[1]

(b) Diagram **C** shows the Rutherford model of the atom.

Describe the Rutherford model of the atom.

....

....

....
[2]

Self-assessment and reflective learning page
Materials and solutions

Tick (✓) the box which best shows how you feel about each statement.

Statement	I don't know	I need more practice	I understand
I can describe the Rutherford model of the structure of an atom.			
I know that electrons have negative charge, protons have positive charge and neutrons have no charge.			
I know that the electrostatic attraction between positive and negative charge is what holds together individual atoms.			
I know that purity is a way to describe how much of a specific chemical is in a mixture.			
I understand that the concentration of a solution relates to how many particles of the solute are present in a volume of the solvent.			
I can describe how paper chromatography can be used to separate and identify substances in a sample.			
Thinking and working scientifically			
I can use symbols to represent scientific ideas.			
I can decide what equipment is required to carry out an investigation or experiment and describe how to use it appropriately.			
I can make conclusions by interpreting results.			
I can evaluate experiments and investigations, and suggest improvements, explaining any proposed changes.			

Mark for end of unit test _____ /20 marks

What went well in this topic?

What could you do better next time? What parts of the course could your teacher go through in a revision lesson which would support your improvement?

Teacher comment:

End of Unit Test: Chemical changes
Total = 20 marks

Name: .... Class: ....

Date: ....

1 Some elements are very reactive, whereas other elements do not react at all.

   (a) Xenon is a noble gas. It is very unreactive.

   Circle the word that means unreactive.

   explosive   **unstable**   **inert**   **flammable**

   [1]

   (b) A teacher shows her class the reaction of three metals: sodium, potassium and gold, with water.

   (i) Write down **one** safety precaution the teacher should take when carrying out this experiment.

   ....

   ....
   [1]

   (ii) When the teacher adds sodium to water, it fizzes, forms a ball and floats on top of the water.

   When the teacher adds potassium to water, it fizzes a lot, floats and burns with a lilac flame.

   When gold is added to water, there is no reaction.

   Place the three metals sodium, potassium and gold in order from the most to the least reactive.

   most reactive ....

   ....

   least reactive ....

   [1]

2 Kwame completes a series of experiments. He has four salts labelled **A, B, C** and **D**.

   He adds each salt to a separate beaker of water.

   He measures the temperature of the water **before** and **after** the salt has been added.

   (a) Complete the table to show the results of the experiment.

Salt	Temperature at the start (°C)	Temperature at the end (°C)	Temperature change (°C)
A	20.0	23.5	+3.5
B	18.0	25.0	....
C	....	22.0	−1.0
D	19.0	23.0	....

   [1]

   (b) The temperature change for salt **A** is +3.5 °C. Explain why there is a **plus** sign in front of the value 3.5.

   ....

   ....
   [1]

3 Ricardo investigates the reaction of oxygen with four metals: magnesium, zinc, iron and copper.

   (a) He uses some tongs to hold a small piece of copper metal in a Bunsen burner flame.

   Name the compound formed when copper is burnt in air.

   ....
   [1]

   (b) When Ricardo heats a piece of magnesium it burns with a brilliant white flame and makes a new compound called magnesium oxide.

   Write down a word equation for this reaction.

   ....
   [2]

4 Ashani injures her knee playing football. She applies a chemical cold pack to the knee to reduce the swelling.

   (a) Give the name of the type of chemical reaction that takes place in the cold pack.

   ....
   [1]

   (b) Explain your answer.

   ....

   ....
   [1]

5 Metals can be placed into a reactivity series.

   most reactive — potassium

   calcium

   magnesium

   least reactive — silver

   A teacher reacts magnesium with hydrochloric acid. She finds that the temperature of the mixture increases by 2 °C and that bubbles form.

   (a) Name the salt made in this reaction.

   ....
   [1]

   (b) Predict what would be seen when silver is added to hydrochloric acid.

   ....

   ....
   [1]

   (c) Predict what would be seen when calcium is added to hydrochloric acid.

   ....

   ....
   [1]

(d) Explain why it would **not** be safe to add potassium to hydrochloric acid in a school laboratory.

....

....
[1]

6 Substances can be pure or impure.

When chlorine is reacted with methane, several compounds are made. The new compounds include chloromethane and dichloromethane.

Draw a line from each **substance** to indicate whether it is **pure or impure**.

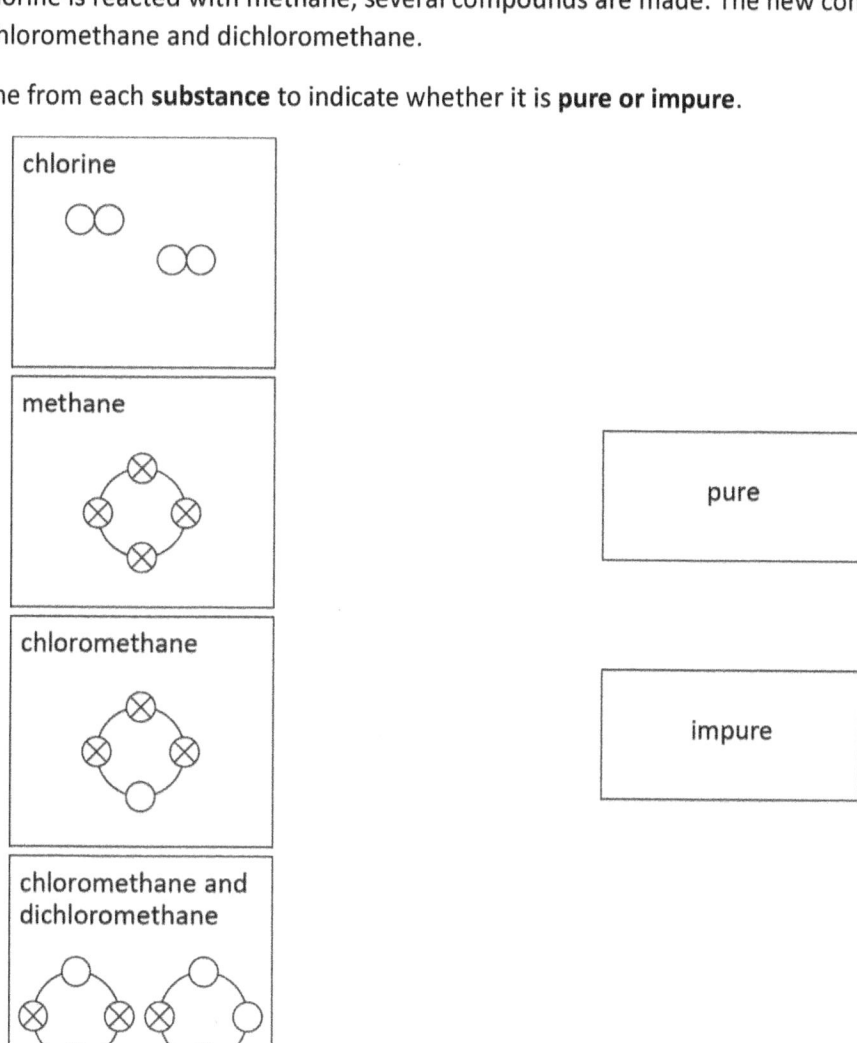

[2]

7 Aluminium nitrate is a salt that is soluble in water.

The table below shows the mass of the salt aluminium nitrate that dissolves in 100 cm³ of water at different temperatures.

Temperature (°C)	0	10	20	30	40	50	60	70	80	90	100
Mass of aluminium nitrate (g)	60	67	74	82	89	96	106	120	132	153	160

(a) Deepa predicts that about 170 g of aluminium nitrate will dissolve in 100 cm³ of water at 110 °C.

Write down why it is not possible to determine the solubility of aluminium nitrate in water at 110 °C in the school laboratory.

....
[1]

(b) Complete the graph to show the results for aluminium nitrate.

   (i) Plot the missing points.

[1]

   (ii) Add a line of best fit.

[1]

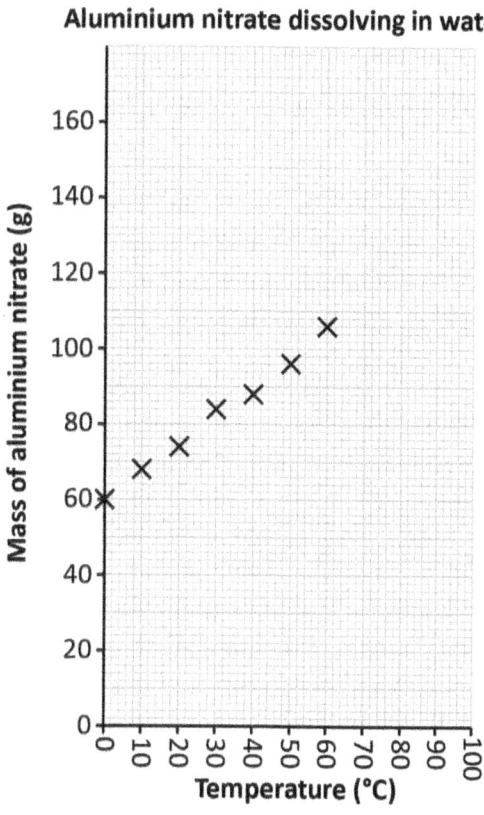

Aluminium nitrate dissolving in water

(c) Describe how the solubility of aluminium nitrate changes as the temperature increases.

....

....

[1]

Self-assessment and reflective learning page
Chemical changes

Tick (✓) the box which best shows how you feel about each statement.

Statement	I don't know	I need more practice	I understand
I can use word equations to describe reactions.			
I can describe the reactivity of metals with water and dilute acids.			
I know that some processes and reactions are endothermic or exothermic and that this can be identified by temperature change.			
I know that reactions do not always lead to a single pure product and that sometimes a reaction will produce an impure mixture of products.			
I can describe how the solubility of different salts varies with temperature.			
I understand that some substances are generally unreactive and can be described as inert.			
Thinking and working scientifically			
I can identify whether a given hypothesis is testable.			
I can make predictions of likely outcomes for a scientific inquiry based on scientific knowledge and understanding.			
I can describe how to carry out practical work safely.			
I can describe trends and patterns in results.			
I can make conclusions by interpreting results.			
I can present and interpret observations and measurements appropriately.			

Mark for end of unit test _____ /20 marks

What went well in this topic?

What could you do better next time? What parts of the course could your teacher go through in a revision lesson which would support your improvement?

Teacher comment:

End of Unit Test: Motion
Total = 20 marks

Name: .... Class: ....

Date: ....

1 This question is about speed.

   (a) Complete the equation for speed.

   $$\text{average speed} = \frac{\text{total .... travelled}}{\text{total .... taken}}$$

   [2]

   (b) Identify the correct unit for speed.

   A m   **B** s

   C m/s   **D** m s

   ....
   [1]

2 The graph shows a cyclist on a journey.

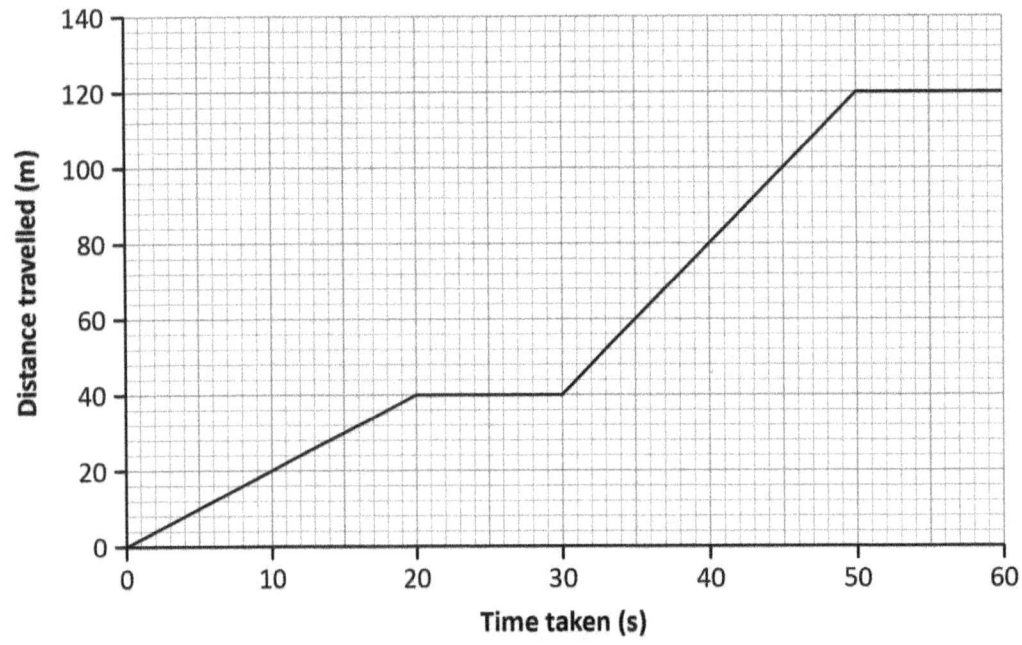

   (a) Write down what the gradient of this graph shows.

   ....
   [1]

38

(b) Which phrase completes the sentence? Tick (✓) the correct box.

Between 20 and 30 seconds, the cyclist was ....

stationary ☐

moving at a constant speed ☐

speeding up ☐

slowing down ☐

[1]

(c) Identify the part of the journey where the cyclist was travelling fastest.

Between .... and ....seconds.

[1]

3 Yang Yi investigates the speed of a marble rolled along a table.

She has a 2 m measuring tape and a choice of timing equipment.

(a) From the list, select the appropriate timing equipment. Circle **one** item.

wall clock   **stopwatch**   **egg timer**

[1]

(b) (i) The diagram shows one result from the investigation.

Write down the distance the marble rolled, to the nearest 0.5 cm.

....cm
[1]

(ii) It took 4.0 s for the marble to roll to this distance.

Calculate the average speed of the marble, in cm/s.

....cm/s
[2]

39

4 Lily investigates the movement of a runner on a track.

She measures the distance the runner moves every 10 seconds for 1 minute.

Her results are shown in the table.

Time (s)	Distance (m)
0	0
10	40
20	80
30	110
40	135
50	150
60	160

Lily has started to draw a graph of her results.

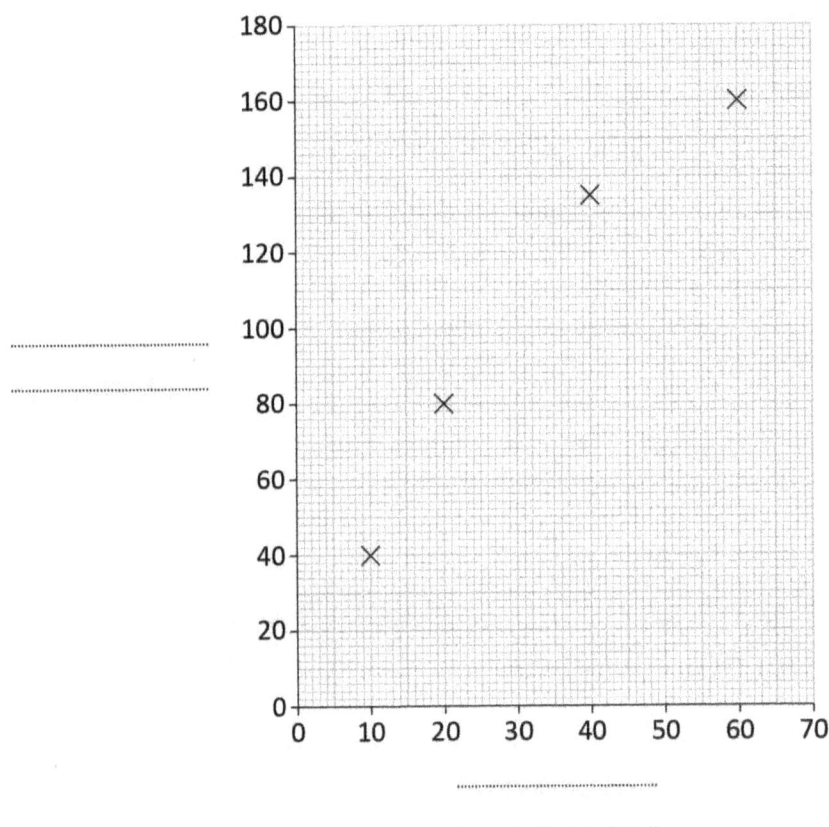

(a) Label the axes of the graph including the correct units.

[2]

(b) Complete plotting the values on the graph.

[1]

(c) The graph is not a straight line. Explain what this shows.

....

[1]

5 The distance–time graphs show the movements of three different animals.

Draw a line to match each **animal** to the correct **distance–time graph**.

[2]

Animal

Graph

A

whale
moves slowly

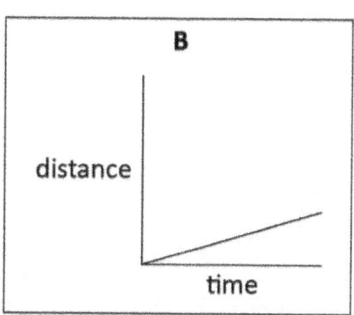

B

hummingbird
moves between flowers to feed

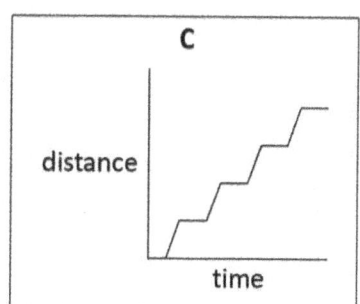

C

cheetah
stalks and chases prey

41

6 Hiro investigates a toy car that rolls down a ramp.

The diagram shows the equipment.

The computer calculates the speed of the car as it passes through the light gate.

The table shows Hiro's results.

Measurement	Speed of car (cm/s)
1	14.0
2	13.0
3	15.0
4	11.0
5	14.0

(a) Identify the anomalous measurement.

speed = ...cm/s
[1]

(b) Hiro determines the average speed as follows:

$$\text{average speed} = \frac{14.0 + 13.0 + 15.0 + 11.0 + 14.0}{5} = 13.4 \text{ cm/s}$$

This is **not** the correct average speed. Explain why.

....
[1]

(c) Calculate the correct average speed.

.................................cm/s
[2]

Self-assessment and reflective learning page
Motion

Tick (✓) the box which best shows how you feel about each statement.

Statement	I don't know	I need more practice	I understand
I can calculate speed using the equation $speed = \dfrac{distance}{time}$.			
I can draw simple distance–time graphs.			
I can interpret distance–time graphs.			
Thinking and working scientifically			
I can use formulae to represent scientific ideas.			
I can decide what equipment is needed to carry out an investigation.			
I can take appropriately accurate and precise measurements.			
I can identify anomalous results.			
I can describe trends and patterns in results.			
I can make conclusions by interpreting results.			
I can present and interpret observations and measurements appropriately.			

Mark for end of unit test _____ /20 marks

What went well in this topic?

What could you do better next time? What parts of the course could your teacher go through in a revision lesson which would support your improvement?

Teacher comment:

End of Unit Test: Forces
Total = 20 marks

Name: .... Class: ....
Date: ....

1 Use the words in the box to answer the questions.

balanced	unbalanced	falls at constant speed	speeds up
		slows down	

(a) The diagram shows a ball that has been dropped.

↓ weight of ball

Complete the sentence.

The ball ....because its weight is ....

[1]

(b) The diagram shows the ball during its fall, when air resistance equals the weight.

↑ air resistance
↓ weight of ball

Complete the sentence.

The ball ....because its weight is ....

[1]

2 This question is about gases.

   (a) Circle the correct terms.

   The particles in a gas are **far apart / close together** and **stay still / move around**.

   [1]

   (b) Kira pumps air into a balloon.

   Tick (✓) the force that causes the balloon to get bigger.

   air resistance ☐

   moment ☐

   normal reaction ☐

   pressure ☐

   [1]

3 Simran uses a spanner to turn a nut.

   (a) Label the pivot and the effort.

   [1]

   (b) Write down **two** ways that the turning force (moment) can be increased.

   1...

   2...

   [2]

4   This question is about forces.

(a) The diagram shows a lamp attached to a wire. The lamp does not move.

weight

Add a force arrow to the diagram to show why the lamp does not move.

[1]

(b) The diagram shows two forces acting on a car.

200 N   130 N

Calculate the resultant force and state its direction.

....
[1]

5 Ashani uses coloured balls to make a particle model of two gases.

She puts black balls at the bottom of one beaker, and white balls in another beaker.

She then puts the two beakers together and shakes them.

The diagrams show what happens.

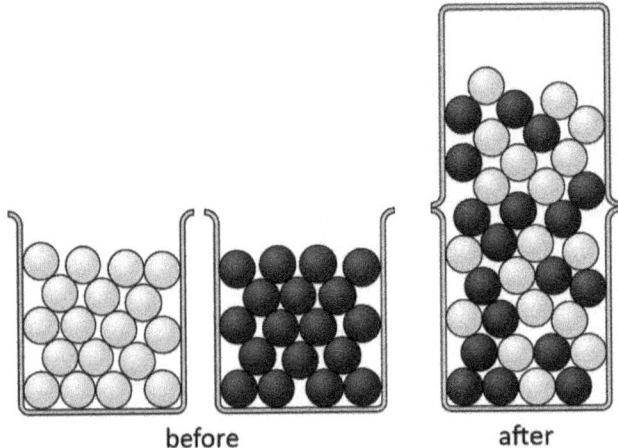

before   after

(a) Identify the force or process that Ashani is trying to model. Circle one term from the word box.

```
air resistance

diffusion

pressure

upthrust
```
[1]

(b) Explain a limitation of this model.

....

....

[1]

6 The image shows a penguin on an iceberg floating on the sea.

(a) Name the force that stops the penguin and ice from sinking.

....

[1]

(b) The iceberg weighs 3700 N and the penguin weighs 400 N.

Calculate the force needed to stop the penguin and iceberg sinking.

....

[1]

7 Carlos investigates turning forces using a lever balance.

He sets up the situation in the diagram, and the lever balances.

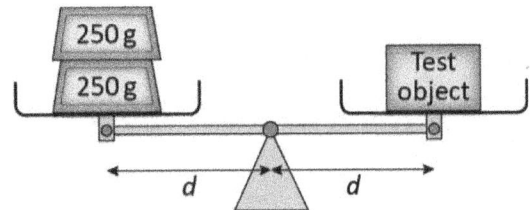

(a) Describe how Carlos could make the results of his investigation more reliable.

....

....
[1]

(b) Predict what will happen in the situations shown in the diagrams.

(i)

Prediction: ....

[1]

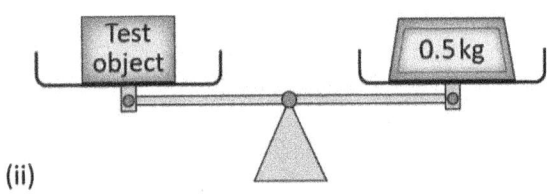

(ii)

Prediction: ....

[1]

8 Lorna measures the weight of cubes X, Y and Z.

She also calculates the area of one face of each of the cubes.

Lorna's measurements are shown in the table.

Cube	Weight (N)	Area of face (m^2)	Pressure cube exerts on table (N/m^2)
X	10	0.10	100
Y	20	0.10	
Z	30	0.20	

Lorna predicts that cube X will exert the largest pressure when placed on a desk.

(a) Write down the equation to calculate pressure.

....
[1]

(b) Calculate the pressure exerted by cube Y and cube Z. Write your answers in the table.

[2]

(c) Evaluate Lorna's prediction.

....

....
[1]

Self-assessment and reflective learning page
Forces

Tick (✓) the box which best shows how you feel about each statement.

Statement	I don't know	I need more practice	I understand
I can describe the effects of balanced and unbalanced forces on motion.			
I can identify turning forces.			
I can calculate turning forces using the equation moment = force × distance.			
I can explain that pressure is caused by a force applied over an area.			
I can calculate pressures using the equation $\text{pressure} = \frac{\text{force}}{\text{area}}$.			
I can use the particle model to explain pressure in gases and liquids.			
I can describe diffusion as the mixing of substances by the movement of particles.			
Thinking and working scientifically			
I can describe what an analogy is and how it can be used as a model.			
I can use symbols and formulae to represent scientific ideas.			
I can make predictions based on scientific knowledge and understanding.			
I can evaluate whether measurements have been repeated sufficiently to be reliable.			
I can describe the accuracy of predictions and suggest why they were or were not accurate.			

Mark for end of unit test _____ /20 marks

What went well in this topic?

What could you do better next time? What parts of the course could your teacher go through in a revision lesson which would support your improvement?

Teacher comment:

End of Unit Test: Light and magnets
Total = 20 marks

Name: .... Class: ....

Date: ....

1 The diagram shows a ray of light reflected from a plane mirror.

(a) Add labels to the diagram for the **angle of incidence** and the **angle of reflection**.

[1]

(b) Write down the law of reflection.

....

[1]

2 This question is about refraction.

(a) Complete the definition. Use words from the word box.

reflection	bending	transparent	opaque

Refraction is the .... of light as it travels from one

.... substance to another.

[1]

(b) Identify the properties of a light ray that change during refraction. Tick (✓) **two** boxes.

brightness ☐

colour ☐

direction ☐

speed ☐

[1]

3 The diagram shows white light incident on a prism.

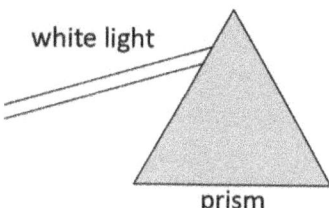

(a) Describe how the white light changes when it leaves the prism.

....

[1]

(b) Write down the name of this process.

....

[1]

4 Damin operates the lights for a school play.

Damin shines a cyan-coloured light through different coloured filters onto the stage.

Cyan is a mixture of blue and green light.

(a) When Damin shines the light onto a blue filter, the stage appears blue. Explain why this happens.

....

....

[1]

(b) Explain what will happen if Damin shines the cyan light onto a red filter.

....

....

[1]

5 This question is about magnetic fields.

(a) Omar investigates how two magnets behave when they are placed near to each other.

Complete the table.

Poles brought close together	Attract or repel?
S N	(i)..
S S	(ii)...

[1]

(b) Complete the diagram to show the magnetic field around a bar magnet.

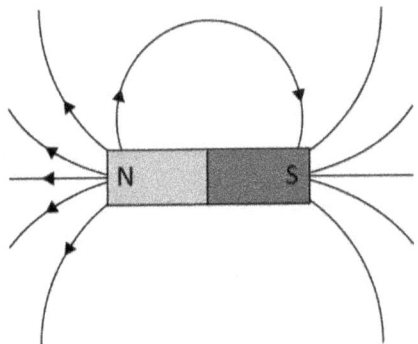

[1]

6 Siani investigates the reflection of light.

She measures the angles of incidence and reflection for a light ray.

The table shows her results.

Angle of incidence	Angle of reflection
20°	20°
40°	40°
60°	
80°	

(a) Complete the table.

[1]

(b) Siani tries to use a crumpled piece of foil in place of the mirror, but she cannot find a bright reflected ray to measure.

Explain what is happening.

....

....
[1]

7 Sandeep investigates colour addition using red, green and blue lights.

He writes a hypothesis:

Shining any three coloured lights onto one area will produce white light.

(a) Sandeep shines a blue light and two red lights onto a white screen.

What colour light will Sandeep produce? Circle the correct answer.

   blue   magenta   red   white

[1]

(b) Rewrite Sandeep's hypothesis to suggest how he can produce white light.

Make sure that your hypothesis is testable using Sandeep's equipment.

....

....

....
[1]

57

8 Tashika wants to investigate and show the strength and direction of the magnetic field around a horseshoe magnet.

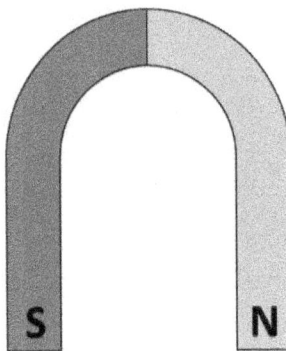

(a) Tashika starts writing an equipment list in the table.

Complete the table.

Equipment
horseshoe magnet
white paper
....
....

[1]

(b) The diagram shows Tashika's results.

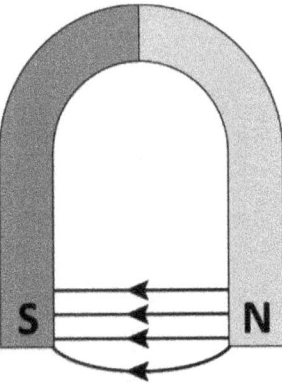

Explain what Tashika needs to do to complete the results.

....

....

[1]

9 Emily has a plastic rod, some copper wire and a battery.

(a) Describe how Emily can make an electromagnet using this equipment.

....

....
[1]

Emily uses steel paperclips to investigate the strength of her electromagnet.

She increases the number of turns in the coil and counts how many paperclips the electromagnet can hold up.

The table shows the results.

Number of turns	Number of paperclips
4	2
8	4
12	6

(b) Use the values in the table to predict how many paperclips will be held up by a coil of 20 turns.

Show your working.

......................................paperclips.
[2]

(c) A crane in a scrapyard uses an electromagnet to pick up steel items and move them around.

Explain why the electromagnet contains an iron core.

....

....
[1]

Self-assessment and reflective learning page
Light and magnets

Tick (✓) the box which best shows how you feel about each statement.

Statement	I don't know	I need more practice	I understand
I can describe reflection of light at a plane surface.			
I can use the law of reflection.			
I can describe refraction of light at a boundary in terms of change of speed.			
I know that white light is made of many colours.			
I can use a prism to show dispersion of white light.			
I can describe how colours of light can be added, subtracted, absorbed and reflected.			
I can describe a magnetic field.			
I understand that a magnetic field surrounds a magnet and exerts a force on other magnetic fields.			
I can describe how to make an electromagnet.			
I know that electromagnets have many applications.			
I can investigate factors that change the strength of an electromagnet.			
Thinking and working scientifically			
I can say whether a hypothesis is testable.			
I can describe how scientific hypotheses can be supported or contradicted by evidence.			
I can make predictions based on scientific knowledge and understanding.			
I can plan investigations.			

I can decide what equipment is needed to carry out an investigation.			
I can describe trends and patterns in results.			
I can make conclusions by interpreting results.			

Mark for end of unit test _____/20 marks

What went well in this topic?

What could you do better next time? What parts of the course could your teacher go through in a revision lesson which would support your improvement?

Teacher comment:

# End of Unit Test: The Earth in space
Total = 20 marks

Name: .... Class: ....

Date: ....

1 Ewan states that the Earth has a magnetic field.

   (a) Name the device Ewan could use to show this magnetic field.

   ....

   [1]

   (b) Identify **two** substances found in the Earth's core that produce this magnetic field.

   1...

   2...

   [1]

2 This question is about resources.

   (a) Draw a line to match each **resource** with one of its **uses**.

Resource	Use
rice	refined to make plastics
crude oil	as food
hot rocks (geothermal energy)	to heat homes

   [2]

   (b) Write down the resource(s) in **(a)** that is/are renewable.

   ....

   [1]

3 (a) Complete the sentence about the Earth's temperature. Circle the correct word.

The average temperature on Earth **stays the same / changes** over long periods of time.

[1]

The graph shows estimates of Earth's average temperature between 420 000 and 320 000 years ago.

The values on the *y*-axis show the temperature compared to now.

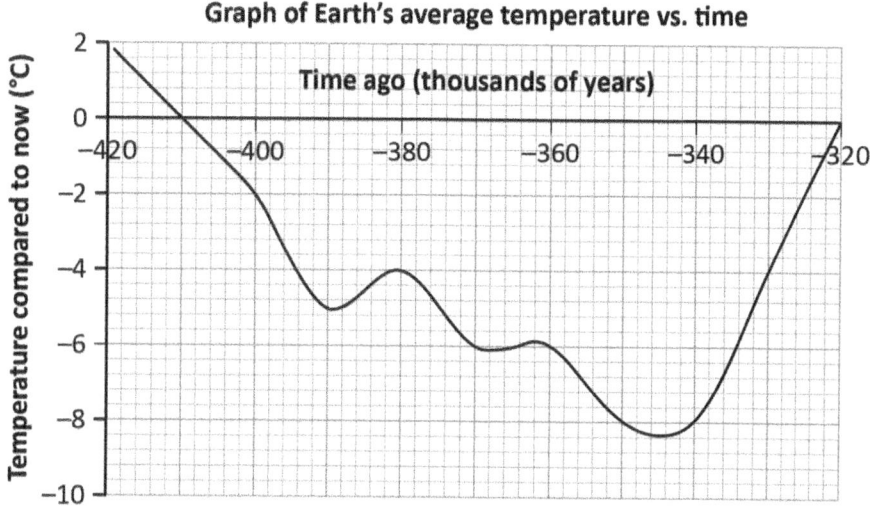

(b) Name the part of the temperature cycle that occurred between about 400 000 and 330 000 years ago. Tick (✓) **one** box.

A glacial period ☐

B global warming ☐

C interglacial period ☐

D warm period ☐

[1]

(c) The average temperature today is about the same as that on two other occasions in the graph.

One occasion was 320 000 years ago.

Write down the other occasion.

....years ago.

[1]

(d) Explain how the places where animal fossils and skeletons are found can help scientists to estimate the temperature thousands of years ago.

....

....
[1]

4 This question is about our galaxy, the Milky Way.

(a) The table lists some types of objects found in the Milky Way.

Type of object	Order from smallest (1) to largest (5)
asteroid	....
black hole	....
interstellar dust cloud	....
planet	2
star	....

Write a number from **1** to **5** in the table to sort the objects in order from smallest (**1**) to largest (**5**).

One has been done for you.
[2]

(b) Explain why asteroids do not have atmospheres.

....
[1]

(c) Three astronomers estimated the number of stars in the Milky Way.

The table shows their results.

Astronomer	Estimate
1	200 billion
2	250 billion
3	180 billion

Calculate the average of the estimates. Show your working.

....billion

[2]

5 The graph shows the amount of carbon dioxide in the atmosphere since AD1750.

(a) Explain how the increase in carbon dioxide is affecting the Earth's climate.

....

....

[2]

(b) Explain why it is important that many different teams of scientists around the world have found the same results.

....

[1]

6 Hannah measures the temperature and rainfall in the school grounds each day for one week.

The table shows her results.

Temperature at midday (°C)	21	22	17	16	19	23	22
Rainfall over 24 hours (mm)	0	0	6	41	15	0	4

Hannah writes this conclusion:

The climate around the school changes each day.

(a) Rewrite the conclusion using the correct term.

....

[1]

(b) Explain **two** ways that the investigation would need to change in order to measure climate.

....

....

[2]

Self-assessment and reflective learning page
The Earth in space

Tick (✓) the box which best shows how you feel about each statement.

Statement	I don't know	I need more practice	I understand
I know that the Earth has a magnetic field because the core acts as a magnet.			
I can identify renewable and non-renewable resources and describe how people use them.			
I understand that there is evidence that the Earth's climate has a cycle of warm periods and ice ages that take place over long time periods.			
I understand that the Earth's climate can change due to changes in the atmosphere.			
I can describe the difference between climate and weather.			
I can describe a galaxy in terms of interstellar dust and gas, stars and planetary systems.			
I can describe asteroids as rocks that are smaller than planets.			
I can describe how asteroids formed.			
Thinking and working scientifically			
I can sort, group and classify objects.			
I can evaluate whether measurements have been repeated enough to be reliable.			
I can identify trends and patterns in results.			
I can make conclusions by interpreting results.			
I can explain the limitations of conclusions.			

I can evaluate investigations and suggest improvements.			
I can interpret measurements appropriately.			

Mark for end of unit test _____ /20 marks

What went well in this topic?

What could you do better next time? What parts of the course could your teacher go through in a revision lesson which would support your improvement?

Teacher comment:

End of Year Test 1
Total = 50 marks

Name: .... Class: ....

Date: ....

1 This question is about speed.

(a) Which of the following is the correct formula for speed? Tick (✓) **one** box.

speed = distance × time ☐

$speed = \dfrac{distance}{time}$ ☐

$speed = \dfrac{time}{distance}$ ☐

$speed = \dfrac{distance}{(time)^2}$ ☐

[1]

(b) A cyclist travels 400 metres in 20.0 seconds.

Calculate the cyclist's speed. Include the correct units in your answer.

....

[2]

2 The diagram shows Rutherford's model of an atom.

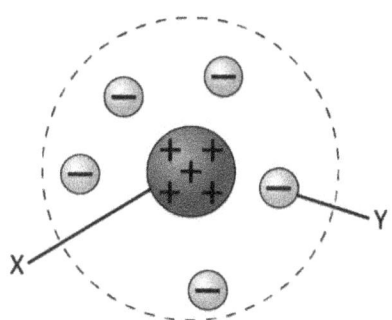

(a) Name the parts labelled X and Y.

X: ....

Y: ....

[2]

(b) Complete the sentences.

The particles labelled **Y** are kept in orbit of **X** because they have opposite

....

Most of the space in an atom is ....

[2]

3 Chen is ill because he has been infected by a microorganism.

(a) Name the cells in the human body which destroy microorganisms that cause infections.

....

[1]

(b) Although he is ill, Chen still needs nutrients.

Name the part of the blood that carries nutrients to Chen's cells.

....

[1]

4 Galaxies contain interstellar gas and dust.

(a) Name the **two** gases that make up nearly all interstellar gas.

.... and ....

[2]

(b) Large amounts of the elements carbon, oxygen and silicon are present on Earth.

Suggest how these elements may have come to Earth from interstellar dust.

....

[1]

5 Tashika investigates how much magnesium chloride will dissolve in tap water at different temperatures.

The table shows her results.

Column 1 Temperature (°C)	Column 2 Mass of magnesium chloride that dissolves (g)
1	53.0
20	54.4
40	57.5
60	61.0

(a) Name the physical property of magnesium chloride that Tashika has measured in **Column 2**.

....
[1]

(b) Write down a variable that must be kept constant (the same) for this investigation.

....
[1]

(c) Predict the mass of magnesium chloride that would dissolve if the temperature was 80 °C.

Circle **one** answer.

57.5　　　　　　　　61.0　　　　　　　　64.5

[1]

(d) Explain your answer to (c). Include the word 'particles' in your answer.

....

....
[1]

6 This question is about balanced diets.

(a) Draw a line to match each **nutrient** with its **function**.

Nutrient	Function
carbohydrates	carrying substances around the body and temperature control
fats	growth and repair
proteins	source of energy
water	store of energy

[2]

(b) Write down another **nutrient** that is missing from part (a).

....

[1]

The table shows some information from a food label.

NUTRITION INFORMATION	
	Amount per 100 g
Energy	1200 kJ
Protein	10.0 g
Carbohydrate	35.0 g
• sugars	20.0 g
• starch	15.0 g
Fat	5.0 g
Fibre	25.0 g

A normal serving of this food is 30 g.

(c) Calculate the mass of sugars in one normal serving. Show your working.

.................................g

[2]

7 Anna investigates the pressure caused on the desk by objects with different bases.

Each object weighs 10 N.

The table shows Anna's results.

Object	Weight (N)	Area of base (m^2)	Pressure (N m)
A	10	0.1	100
B	10	0.2	....
C	10	0.4	....

(a) The table headings contain a mistake.

Identify what is incorrect.

....

[1]

(b) Write the correct heading.

....

[1]

(c) The formula for pressure is:

$$\text{pressure} = \frac{\text{force}}{\text{area}}$$

Calculate the missing values and write them in the table.

[2]

(d) The pressure from the objects arises because of the contact between the object and the desk.

Explain briefly how a gas exerts pressure on the walls of its container.

....

....

[1]

8 The diagram shows a plant cell.

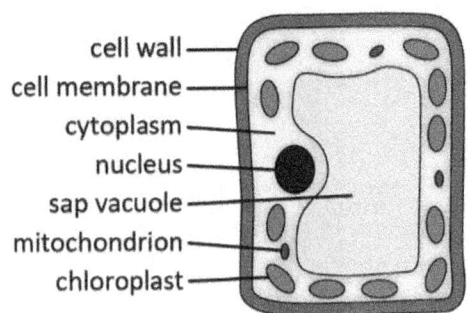

(a) Complete the following word equation for aerobic respiration.

oxygen + ... → water + ...

[2]

(b) Name the part of the cell where aerobic respiration takes place.

....

[1]

9 There are many pieces of evidence that show Earth's climate is changing.

(a) Scientists drill into glaciers and take out long ice cores.

The cores contain frozen bubbles of gas.

Explain how studying these bubbles provides evidence of changes in the Earth's atmosphere.

....

....

[1]

(b) Scientists study fossils and rocks to estimate the sea level long ago.

Use the graph to answer the questions that follow.

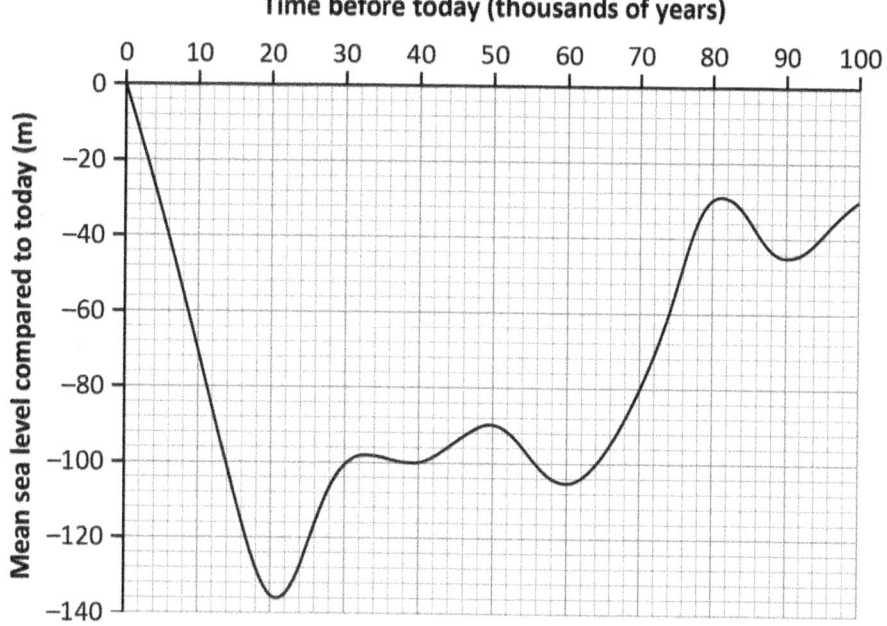

(i) Identify when the sea level was **lowest** in the past 100 000 years.

....

[1]

(ii) The sea level is lowest in a glacial period. Explain why.

....

....

[1]

10 Harry completes a field survey of some wild plants in a public park. He took ten samples.

The table shows Harry's results.

Type of plant	Sample number									
	1	2	3	4	5	6	7	8	9	10
ivy		✓	✓		✓	✓		✓		
honeysuckle	✓	✓		✓						✓
knotweed						✓	✓			

(a) Calculate the percentage of samples that contained knotweed. Show your working.

.....................................%
[2]

(b) Knotweed is an invasive species.

Knotweed is controlled because it can cause a lot of damage to an ecosystem.

Describe how an invasive species can damage an ecosystem.

....

....
[1]

11 Lodestone is a magnetic rock.

Sandeep investigates the properties of lodestone.

The diagram shows the equipment.

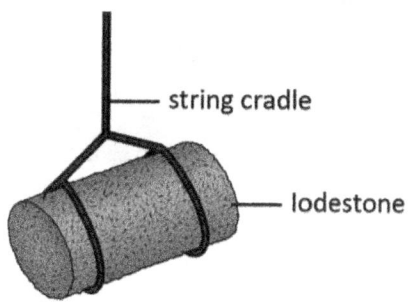

(a) Sandeep brings each of the objects listed in the table below close to the lodestone.

Complete the table with ticks (✓) to predict the results.

Object	Effect of lodestone on object		
	No effect	Attracts	Repels
iron filings	☐	☐	☐
wooden ruler	☐	☐	☐
steel paperclip	☐	☐	☐
table tennis ball	☐	☐	☐

[2]

(b) Write down which of Sandeep's objects could be used to show the magnetic field lines of the lodestone.

....

[1]

12 Kemi has made a model of a hinge joint.

Look at the diagram.

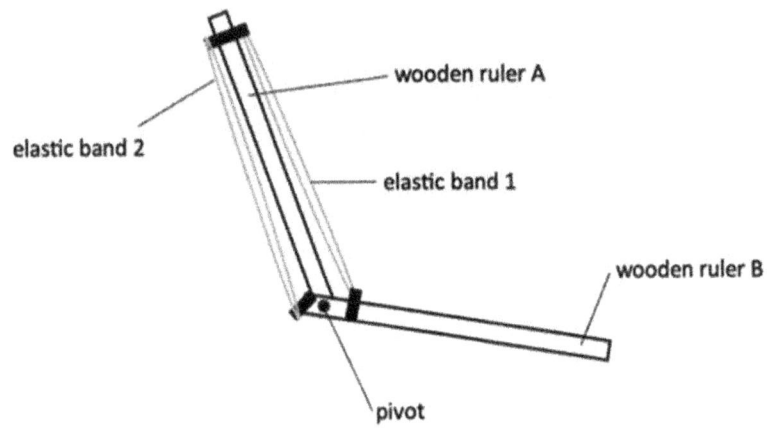

(a) Write down which parts of the model represent bones and which parts represent muscles.

bones: ....

muscles: ....

[1]

(b) Describe how this model could be used to show muscles working as antagonistic pairs.

....

....

....

....

[2]

13 Demar investigates coloured light.

He has this equipment:

3 white lights
2 red filters
2 green filters.

(a) Demar shines white light through a green filter.

What colour of light will this produce?

....

[1]

(b) Demar wants to make a model of a traffic light.

He can only use the equipment from the list at the start of question 13.

He needs to produce three coloured lights: red, yellow and green.

Explain how Demar can do this.

....

....

....

....

[2]

14 Mika investigates the reaction of two solutions: sodium hydrogen carbonate and citric acid.

Mika:

- i. measured the temperature of the solutions at the start of the experiment (they were the same)
- ii. mixed the two solutions together
- iii. waited 10 seconds
- iv. measured the temperature of the mixed solutions.

The table shows the results.

Test	Temperature (°C)		
	At start	After 10 seconds	Change
A	22	17	−5
B	21	18	−3
C	26	16	−10
D	22	18	−4

(a) Identify any anomalous results. Write down the letter or letters of the test(s).

....

[1]

(b) Calculate the average temperature change. Show your working.

Average change = ...°C
[2]

(c) Is this reaction endothermic or exothermic?

....

[1]

(d) Suggest **one** way that Mika could reduce the number of anomalous results.

....

....
[1]

(e) Sports injuries can be treated using packs that contain the two solutions in Mika's experiment.

A pack is placed on a painful, damaged muscle.

Describe how the pack affects the muscle.

....

....
[1]

End of Year Test 2

Total = 50 marks

Name: .... Class: ....
Date: ....

1   The diagram shows a joint in the human body.

(a) Identify the type of joint shown. Circle **one** answer.

    ball and socket   hinge   see-saw

[1]

(b) Describe the function of the part of the joint labelled X.

....

[1]

2   The table shows data about different types of objects P, Q and R, found in space.

Object	Mainly made of rock?	Has an atmosphere?
P	✓	
Q		✓
R	✓	✓

(a) One of the objects is an asteroid. Write down the letter of this object.

....

[1]

(b) Most of the asteroids in our solar system orbit the Sun in the 'asteroid belt'.

This 'belt' is found between the orbits of two planets. Name these two planets.

....and ....

[1]

3 Keiko investigates electromagnets.

She has a long, straight piece of copper wire and a hollow cardboard cylinder.

(a) Write a sentence describing how Keiko could make an electromagnet using this equipment.

....

[1]

(b) Keiko investigates the number of paperclips that her electromagnet can lift off a table.

She tests this using different materials for the core of her electromagnet.

Her results are in the table.

Material	Number of paperclips lifted up
aluminium	5
cardboard	5
iron	8
plastic	4

Write down the material that works best as the core of an electromagnet.

....

[1]

4 The diagram shows part of a food chain for a freshwater habitat.

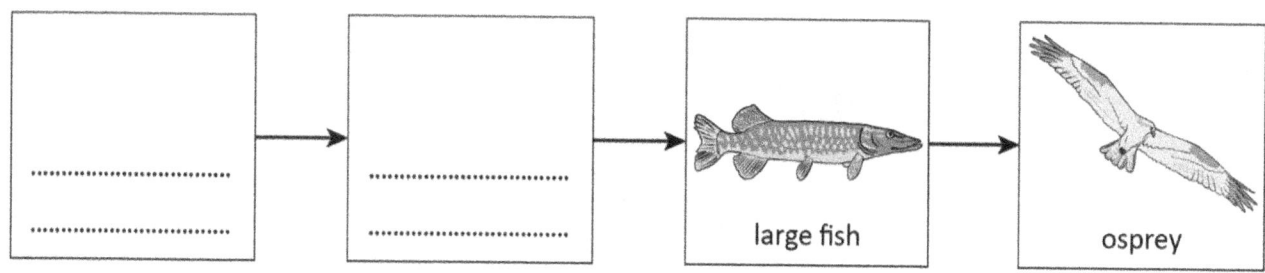

(a) Complete the diagram using **two** of the species from the word box.

```
elephant        small fish        tiny plants
              large tree
```

[2]

(b) Which organism in your food chain is most at risk from bioaccumulation of pesticides?

....

[1]

5 The table lists sub-atomic particles.

Particle	Charge
electron	negative
....	no charge
proton	....

(a) Complete the table.

[2]

(b) The diagram shows a possible model of an atom.

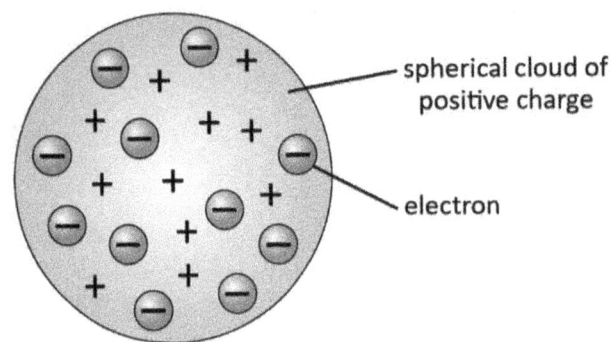

Explain **one** feature of this model that is incorrect.

....

....
[1]

6 The diagram shows two weights arranged on a beam.

The beam rests on a pivot.

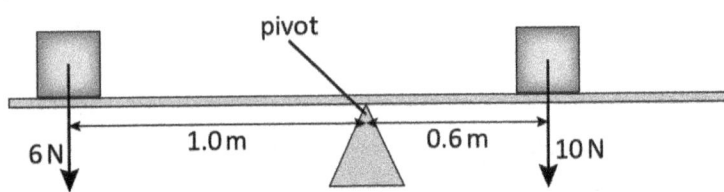

(a) Calculate the moments of the weights.

Remember that: moment = force × distance from the pivot.

anticlockwise moment = ...

clockwise moment = ...
[2]

(b) State whether the beam is balanced or unbalanced.

....

[1]

(c) The weight on the right-hand side of the beam can be moved along the beam.

Predict what will happen if this is moved to the right.

....

[1]

7 A factory is supplied with energy produced by three different resources.

The pie chart shows the share of each resource.

■ natural gas 55%
☐ electricity from diesel generator 13%
▨ electricity from wind turbines ??%

(a) Which of these resources is renewable?

....

[1]

(b) Calculate the percentage of the total energy that is provided by wind turbines. Show your working.

.................................%

[2]

8 Humans need nutrients to live, including proteins, carbohydrates, fats, vitamins, minerals and water.

(a) Which of these nutrients are sources of energy?

....

....
[1]

(b) Energy is released inside cells by aerobic respiration.

Complete the word equation for this reaction.

oxygen + ... → .... + water

[1]

9 Finn investigates reflection of light.

The diagram shows a candle in front of a plane mirror, and the position of Finn's eye.

A light ray from the candle flame is incident on the mirror.

(a) Complete the diagram to include the reflected ray.

[1]

(b) The candle is 50 cm in front of the mirror.

(i) Describe the size and shape of the image Finn will see.

....
[1]

(ii) Describe where the image will appear.

....
[1]

Finn replaces the mirror with a glass block to observe refraction of light.

The diagram shows the arrangement of his equipment.

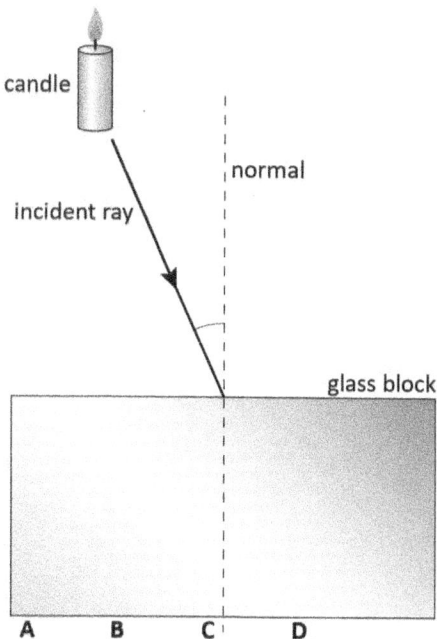

(c) Predict the position of the light ray on the other side of the glass block.

Choose **one** letter from **A, B, C** or **D**.

....

[1]

(d) Finn moves the candle so it is on the normal line.

Predict the new position of the light ray.

Choose **one** letter from **A, B, C** or **D**.

....

[1]

10 Amrit has found two tables of data about temperature.

Table X

Date	Temperature at midday (°C)
11 March	23
12 March	25
13 March	30
14 March	26

Table Y

Year	Average temperature for year (°C)
1990	27.4
2000	27.8
2010	28.1
2020	28.3

(a) Which data show changes in **weather**? Tick (✓) **one** box.

neither table ☐

table **X** only ☐

table **Y** only ☐

both tables **X** and **Y** ☐

[1]

(b) Calculate the average temperature for Table **X**. Show your working.

.....................................°C

[2]

(c) Look at Table **Y**.

Describe the trend in the data.

....

....

[1]

11 Nadiya investigates three different metals, **L, M** and **N**.

She knows that they are calcium, iron and silver, but she does not know which is which.

Nadiya heats each metal in a Bunsen flame to observe its reaction with oxygen from the air.

(a) Write down **two** safety precautions that Nadiya should take.

....

....

[2]

The table shows Nadiya's results.

Metal	With oxygen when heated
L	no reaction, sample remains very shiny
M	burns quickly with red flame
N	dull black layer forms on sample

(b) Write down the metals in order of their reactivity, most reactive first and least reactive last.

...................................is more reactive than ...., which

is more reactive than ....

[1]

(c) Nadiya predicted that **L** is silver because it is very shiny.

Was Nadiya's prediction correct?

Explain your answer using the results from the test.

....

....

[2]

(d) Metal **M** is calcium.

Complete the word equation for the reaction of calcium with oxygen.

calcium + oxygen → ....

[1]

12 Sienna takes part in a race and runs 12 km in 1 hour.

She uses a smart watch to monitor her speed throughout the race.

(a) Write down the definition of a control variable.

....

[1]

(b) Suggest **one** control variable for Sienna's race.

....

[1]

(c) Sienna's watch shows that she travelled at a constant speed for the whole race.

Draw a line on the graph paper below to show how Sienna's distance travelled changed over time.

[2]

13 Ncuti tests the purity of some copper sulphate crystals using chromatography.

The diagram shows the result of the test.

(a) Ncuti concludes that the copper sulfate is not pure.

Explain how the chromatogram shows this.

....

....

[1]

Ncuti compares his chromatogram with those of three other substances.

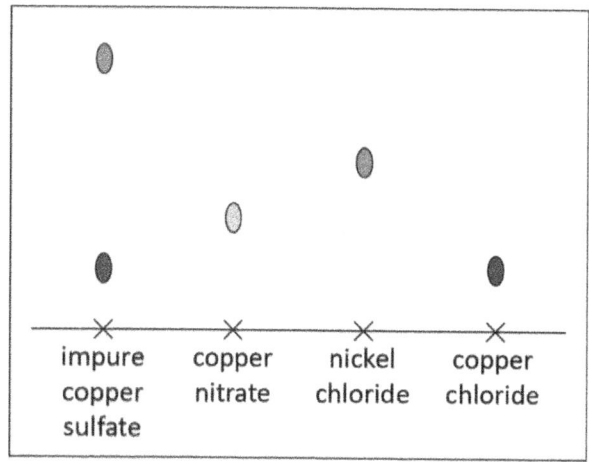

(b) Use the chromatograms to identify the impurity. Circle the substance.

copper nitrate   nickel chloride   copper chloride

[1]

(c) Describe and explain **one** other test that Ncuti could use to show that the copper sulfate is not pure.

....

....

[2]

14 Read the following statement about the respiratory system.

Movements of muscles in your respiratory system that cause air to move in and out of your lungs.

(a) Which term is defined in this way? Tick (✓) **one** answer.

breathing ☐

diffusion ☐

gas exchange ☐

respiration ☐

[1]

(b) The diagram shows the main parts of the respiratory system.

ribcage falls

diaphragm

(i) What process does this diagram show? Circle **one** answer.

exhaling   inhaling

[1]

(ii) Look at the diaphragm on the diagram.

Complete the text label and add an arrow to show how the diaphragm moves.

[1]

(c) Yusuf investigates information available online about the effects of smoking.

He finds this graph, which shows the mean capacity (volume) of the lungs for people who live in a small town.

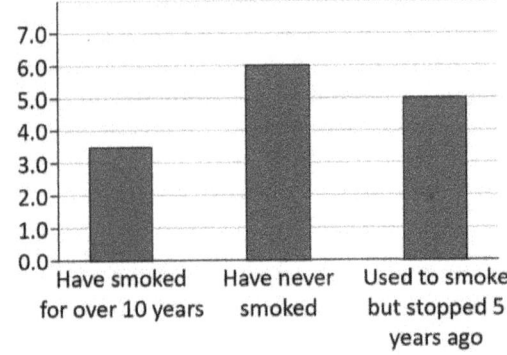

(i) Write a conclusion that can be made from the data in this graph.

....

....

[2]

(ii) Yusuf's teacher tells him to find more information to compare with his conclusion.

Explain why Yusuf needs to do this.

....

[1]

Respiration and movement

Qu	Answer	Mark	Further Information	Spec/ TWS/AO
1(a)	hinge	1		8Bs.01 A01
1(b)		2	One mark for each correct row.	8Bs.01 A02
	Muscle / Bending the leg / Straightening the leg			
	quadriceps / relaxes / contracts			
	hamstrings / contracts / relaxes			
2	plasma – transports cells, nutrients, and carbon dioxide

red blood cell – transports oxygen

white blood cell – protects against pathogens and fights disease | 2 | One mark for two correct lines. Two marks for all three lines correct. | 8Bs.02 A01 |
3(a)	carbon dioxide	1		8Bs.03 A01
3(b)	diffusion	1		8Bs.04 A01
3(c)	It is only one cell thick so gases can diffuse across easily / quickly.	1		8Bs.03 A02

Qu	Answer	Mark	Further Information	Spec/ TWS/AO
3(d)	It flattens / moves down to increase the volume of the chest cavity	1		8Bs.03 AO1
	so the air pressure decreases and air moves into the lungs.	1		
4(a)	mitochondria	1		8Bp.04 AO1
4(b)	glucose	1	Accept sugar.	8Bp.05 AO1
4(c)	The sugar in the drink provides glucose to meet the demands of the muscle cells	1		8Bp.04 AO2
	so they can respire to make energy to contract and enable the runner to keep moving.	1		
5(a)	150	1		8TWSa.02 AO3
5(b)	The heart rate increases because the muscle cells needed to skip are contracting / relaxing and therefore, need more glucose and oxygen for respiration / to produce energy and so, the heart has to pump the blood faster to meet the increased demand.	1 1 1		8Bp.04, 8TWSa.02 8TWSa.03 AO3
5(c)(i)	Because it is more accurate / precise than taking a pulse.	1	Allow other valid suggestions such as prevents missing a beat.	8TWSc.04 8TWSa.04 AO3
5(c)(ii)	Annika should repeat the experiment at least two more times to produce more reliable results.	1		8TWSc.03 8TWSa.04 AO3

Nutrition

Qu	Answer	Mark	Further Information	Spec/TWS/AO
1(a)	carbohydrates and vitamins	2	Can be in either order.	8Bp.01 AO1
1(b)	chicken	1		8Bp.01 AO1
1(c)	A diet that contains all the nutrients in the recommended proportions.	1		8Bp.01 AO1
1(d)	fats and oils – to provide insulation carbohydrates – to provide energy proteins – for growth and repair	2		8Bp.02 AO1
2(a)	Bar drawn to correct height / level at 75% GDA. Bar drawn similar width to other bars with a gap between adjacent bars.	1 1	The data is discontinuous so the bar should be drawn with a gap either side.	8TSWa.05 AO3
2(b)	50%	1		8TSWa.05 AO3
2(c)	40 ÷ 8 = 5 5 kumquats	1		8TSWa.05 AO2
2(d)	To prevent nutrient deficiency diseases such as scurvy.	1		8Bp.01 AO1
3	Imran Because the drink is very high in sugar and contains very low amounts of vitamins and minerals so has little nutritional value.	2	Accept other argument backed up by evidence from label.	8Bp.01 8TWSc.06 AO2
4(a)	A – clamp stand B – tongs	1 1		8TWSc.02 AO3
4(b)	So that Abel can see the mercury column clearly / to avoid reading errors caused by the angle of observation / read the temperature accurately.	1		8TWSc.04 AO3

Qu	Answer	Mark	Further Information	Spec/ TWS/AO
4(c)	safety glasses / goggles	1	Accept tongs.	8TWSc.02 8TWSc.05 AO3
4(d)	digital thermometer	1		8TWSc.02 8TWSa.04 AO3
5(a)	The percentage of males smoking decreased.	1		8TWSa.02 AO3
5(b)	Any one from: cancer, COPD, emphysema, heart disease or high blood pressure.	1		8Bp.03 AO1

Ecosystems

Qu	Answer	Mark	Further Information	Spec/TWS/AO
1(a)	lives physical factors	1 1	One mark each.	8Be.01 AO1
1(b)	rainforest and desert	2	One mark for each ecosystem correctly identified.	8Be.01 AO1
2(a)	pooter	1		8TWSc.02 AO3
2(b)	the conditions in each chamber	1	Also accept descriptions of each chamber condition.	8TWSp.04 AO3
2(c)	Wash his hands thoroughly or wear gloves.	1		8TWSp.05 8TWSc.05 AO3
2(d)(i)(ii)	<table><tr><th rowspan="2">Trial</th><th colspan="4">Number of woodlice</th></tr><tr><th>dark and wet</th><th>dark and dry</th><th>light and wet</th><th>light and dry</th></tr><tr><td>1</td><td>10</td><td>5</td><td>3</td><td>2</td></tr><tr><td>2</td><td></td><td></td><td></td><td></td></tr></table>	3	1 mark for correct heading number of woodlice. 1 mark for correct subheadings describing conditions in each chamber (accept any order). 1 mark for data entered correctly for trial 1.	8TWSc.07 AO3
2(e)	The woodlice prefer the wet and dark conditions.	1		

Qu	Answer	Mark	Further Information	Spec/ TWS/AO
3(a)	One that is non-native to the area and causes damage to the ecosystem.	1		8Be.03 AO1
3(b)	Because they reduced populations of native species in Australia / damaged Australian ecosystems so have become invasive species as well as non-native.	1		8Be.03 AO2
3(c)	The flow of energy from one organism to another.	1		8Be.03 AO1
3(d)	The cane toads would eat more spiders. The cane toads would compete with spiders for food sources that they both eat such as weevils, grasshoppers and beetles.	1 1	One mark for each method.	8Be.03 AO2
4(a)	tuna	1		8Be.02 AO2
4(b)	shark	1		8Be.02 AO2
4(c)	The krill absorb the mercury from the ocean and are eaten by oysters. Because the oysters eat lots of krill the concentration of mercury is higher in the oyster than in the krill. The process continues up the food chain so the shark ends up with the highest concentration of mercury / mercury builds up in the food chain.	2		8Be.02 AO2

Materials and solutions

Qu	Answer	Mark	Further Information	Spec/TWS/AO
1	element compound melting point	2	First two answers can be in either order. Do not allow melting range. Award 2 marks if all three answers correct. Award 1 mark if two answers are correct.	8Cm.04 AO1
2(a)	proton – positive neutron – no charge electron – negative	2	Award two marks if all three answers are correct. Award 1 mark if two answers are correct.	8Cm.02 AO1
2(b)	electrostatic	1	Allow other ways of indicating the answer for example underlining.	8Cm.03 AO1
3(a)	B	1	If no answer is given look at diagram to see if the answer is indicated.	8TWSm.03 8Cp.01 AO2
3(b)	g/cm^3	1	Do not award mark for g.	8Cp.01 AO2
4(a)	4 – Place the chromatography paper to the beaker. 5 – When the solvent reaches near to the top of the paper remove the paper. 6 – Mark where the solvent has reached. 7 – Allow the paper to dry.	2	Award two marks if all four answers are in the correct order. Award 1 mark if any two answers are in the correct order.	8Cp.02 8TWSc.02 AO3
4(b)	When the solvent dries out it will not be visible.	1	Award mark for any reason answer.	8Cp.02 8TWSc.02 AO3
4(c)	It will not run / ink will run.	1		8Cp.02 8TWSc.02 AO3
4(d)(i)	yellow	1		8Cp.02 8TWSa.03 AO3
4(d)(ii)	red and orange	1	Both required but can be in either order.	8Cp.02 8TWSa.03 AO3
4(d)(iii)	It is insoluble (in this solvent).	1	Allow it is NOT a mixture / it is pure / the chromatogram has not been left in the solvent for a long enough time.	8Cp.02 8TWSa.03 AO3

Qu	Answer	Mark	Further Information	Spec/ TWS/AO
4(d)(iv)	Use a different solvent.	1	Accept a named solvent. eg ethanol.	8Cp.02 8TWSa.04 AO3
4(e)	3.0/12.0 = 0.25	1 1	Accept the answer with no working for two marks.	8Cp.02 AO3
5(a)	B	1	If no answer is given look at diagram to see if the answer is indicated.	8Cm.01 AO2
5(b)	(Central) nucleus. Electrons are arranged in a cloud around the nucleus / nucleus has a positive charge.	1 1	Do not allow the electrons are in shells.	8Cm.01 AO1

Chemical changes

Qu	Answer	Mark	Further Information	Spec/ TWS/AO
1(a)	inert	1	Do not allow if more than one answer is circled. Do allow other ways of indicating the answer e.g. underlining.	8Cc.06 AO1
1(b)(i)	safety screen / wearing gloves / using small pieces / having a large container of water / wearing safety goggles	1	Allow other reasonable answers.	8Cc.03 8TWSc.05 AO3
1(b)(ii)	potassium sodium gold	1	All three answers required for the mark.	8Cc.03 AO2
2(a)	+ 7.0 23.0 + 4.0	1	Award mark even if sign is missing. Answers should be given to 1 decimal place.	8Cc.02 8TWSa.05 AO3
2(b)	The temperature goes up / there is a temperature rise / it's exothermic.	1		8Cc.02 8TWSa.03 AO3
3(a)	copper oxide	1	Allow CuO only if formula is completely correct.	8Cc.01 AO1
3(b)	magnesium + oxygen → magnesium oxide	1 1	+ and arrow required for full marks.	8Cc.01 AO1
4(a)	endothermic	1	Do not allow chemical reaction	8Cc.02 AO1
4(b)	The temperature goes down / decreases.	1	Allow energy is transferred from the surroundings to the chemicals.	8Cc.02 AO1
5(a)	magnesium chloride	1	Allow $MgCl_2$ if completely correct.	8Cc.03 AO1
5(b)	no reaction / nothing	1		8Cc.03 8TWSp.03 AO3
5(c)	Temperature rise of greater than 2 °C / lots of bubbles / bubbles are made faster.	1	Allow other reasonable answers.	8Cc.03 8TWSp.03 AO3
5(d)	It is too reactive / too dangerous.	1	Allow other reasonable answers.	8Cc.03 8TWSp.03 AO3

Qu	Answer	Mark	Further Information	Spec/TWS/AO
6	A line connecting the first three materials to pure and bottom material to impure.	2	All four lines required for 2 marks. If two or three lines correct award 1 mark. Do not award mark if more than one line is drawn from a material.	8Cc.04 AO2
7(a)	Water boils / turns into a gas at 100 °C / liquid water cannot get that hot.	1	Allow other reasonable answers.	8TWSp.01 AO3
7(b)(i)	Last four points plotted correctly with crosses or dots.	1	Award mark if there is one mistake. Do not award mark if a very thick pen or pencil is used.	8TWSa.05 AO3
7(b)(ii)	A line of best fit drawn through or very close to the points.	1	Do not award mark if line is sketched. The line must extend at least as far as the first and last points.	8TWSa.05 AO3
7(c)	As the temperature increases the solubility increases.	1	Accept as the temperature decreases the solubility decreases or it's a positive correlation.	8Cc.05 8TWSa.02 AO3

Motion

Qu	Answer	Mark	Further Information	Spec/TWS/AO
1(a)	average speed = $\frac{\text{total distance travelled}}{\text{total time taken}}$	1 1		8Pf.01 8TWSm.03 AO1
1(b)	C: m/s	1		8Pf.01 8TWSm.03 AO1
2(a)	speed	1		8Pf.02 AO1
2(b)	stationary	1		8Pf.02 AO1
2(c)	between **30** and **50** seconds	1	Accept any range between 30 and 50 seconds, e.g. 35 and 45.	8Pf.02 8TWSa.05 AO2
3(a)	stopwatch	1		8Pf.01 8TWSc.02 AO1
3(b)(i)	156.0 cm	1	Decimal place needed for mark as question asks for 'to nearest 0.5 cm'.	8Pf.01 8TWSc.04 AO3
3(b)(ii)	$\frac{156.0}{4.0}$ = 39 cm/s	1 1	accept ECF from 3(b)(ii) accept 39.0 cm/s	8Pf.01 AO2
4(a)	x-axis: Time in seconds y-axis: Distance in metres	1 1	Must include unit for each mark. Accept other units of distance or time for one mark total.	8Pf.02 8TWSa.05 AO1
4(b)	Points plotted at: (30, 110) and (50, 150).	1	Both points needed for one mark. Allow ±(2, 5) range.	8Pf.02 8TWSa.05 AO3
4(c)	speed changes	1	Accept other descriptions. e.g. 'runner slows down' or 'speed reduces'.	8Pf.02 8TWSa.03 AO3
5	whale – Graph B hummingbird – Graph C cheetah – Graph A	2	One or two animals correctly connected to graphs = 1 mark. All three needed for 2 marks.	8Pf.02 AO2
6(a)	11.0	1	Accept '4' if student is indicating measurement 4.	8Pf.01 8TWSa.02 AO2

Qu	Answer	Mark	Further Information	Spec/TWS/AO
6(b)	The anomalous result is included in the calculation.	1	Accept WTTE.	8Pf.01 8TWSa.03 AO3
6(c)	average speed = $$\frac{14.0 + 13.0 + 15.0 + 14.0}{4}$$ = 14.0 cm/s	1 1		8Pf.01 AO2

Forces

Qu	Answer	Mark	Further Information	Spec/TWS/AO
1(a)	The ball **speeds up** because its weight is **unbalanced**.	1	Both correct for one mark.	8Pf.03 AO1
1(b)	The ball **falls at constant speed** because its weight is **balanced**.	1	Both correct for one mark.	8Pf.03 AO1
2(a)	The particles in a gas are **far apart** and **move around**.	1	Both correct for one mark.	8Pf.06 AO1
2(b)	pressure	1		8Pf.06 AO1
3(a)	[Diagram showing wrench with labels: pivot, effort, distance]	1	Both labels correct for one mark.	8Pf.04 AO1
3(b)	Increase the distance (from pivot to effort). Increase the effort.	1 1	Accept 'increase the force'.	8Pf.04 AO1
4(a)	Force arrow added pointing vertically upwards from lamp or on wire, same length as weight.	1	Accept slightly different length. Do not accept different direction.	8Pf.03 8TWSm.03 AO1
4(b)	(200 – 130) = 70 N to the left	1	Value and direction required for one mark. Working not required.	8Pf.03 AO2
5(a)	Diffusion	1		8Pf.07 8TWSm.01 AO2
5(b)	Any one from: not enough 'particles' 'particles' are too close together 'particles' are too large compared to the space they are in 'particles' are not in constant motion.	1	Accept other reasonable limitation.	8Pf.07 8TWSm.01 AO3

6(a)	upthrust	1	Accept pressure.	8Pf.06 AO1
6(b)	(3700 + 400) = 4100 N	1	Accept answer stated without working.	8Pf.03 AO2
7(a)	Take the masses and test object off, then replace them and repeat the measurements.	1	Accept 'repeat the measurements' or similar phrase.	8Pf.04 8TWSc.03 AO3
7(b)(i)	Any one from: tip to the right move downwards on right move upwards on left	1		8Pf.04 8TWSp.03 AO2
7(b)(ii)	balance	1	Accept 'will not tip / move' or similar phrase.	8Pf.04 8TWSp.03 AO2
8(a)	$\text{pressure} = \dfrac{\text{force}}{\text{area}}$	1		8Pf.05 8TWSm.03 AO1
8(b)	Pressure for Y = (20/0.10) = 200 (N/m^2) Pressure for Z = (30/0.20) = 150 (N/m^2)	1 1	Accept each value without working or unit.	8Pf.05 AO2
8(c)	Incorrect, as Y exerts the largest pressure.	1	Must include explanation for one mark.	8Pf.05 8TWSa.01 AO3

Light and magnets

Qu	Answer	Mark	Further information	Spec/ TWS/AO
1(a)	*[Diagram of light ray reflecting off a mirror, showing light ray, angle of incidence, and angle of reflection]*	1	Both correct for one mark.	8Ps.01 AO1
1(b)	angle of incidence = angle of reflection	1	Accept in reverse order.	8Ps.01 AO1
2(a)	Refraction is the **bending** of light as it travels from one **transparent** substance to another.	1	Both correct for one mark.	8Ps.02 AO1
2(b)	direction speed	1	Both correct for one mark.	8Ps.02 AO1
3(a)	Splits / separates into (many) colours.	1		8Ps.03 AO1
3(b)	dispersion	1	Do not accept 'refraction'.	8Ps.03 AO1
4(a)	Blue filter absorbs green light, leaving only blue light.	1		8Ps.04 AO1
4(b)	No light will pass through the filter / stage will turn black because red filter absorbs blue and green light.	1	Must include prediction and explanation for one mark.	8Ps.04, 8TWSp.03 AO2
5(a)	(i) attract (ii) repel	1	Both correct for one mark.	8Pe.01 AO1
5(b)		1	Loop added below magnet and arrows added correctly to all field lines, for one mark.	8Pe.01 AO2

Qu	Answer	Mark	Further Information	Spec/TWS/AO
	(Diagram: bar magnet with S on top and N on bottom, showing magnetic field lines looping from N to S)			
6(a)	Angle of incidence 60°. Angle of reflection **60°**. Angle of incidence 80°. Angle of reflection **80°**.	1	Both correct for one mark.	8Ps.01 AO2
6(b)	Rough / crumpled surface scatters the light.	1	Accept 'reflects light in many directions' or just 'scattering.'	8Ps.01 AO2
7(a)	magenta	1		8Ps.04 8TWSp.02 AO2
7(b)	Shining a red light, a blue light and a green light onto one area will produce white light.	1	Accept any order of the coloured lights.	8Ps.04 8TWSp.01 AO3
8(a)	pen / pencil magnetic compass	1	Both correct for one mark.	8Pe.01 8TWSc.02 AO3
8(b)	Place the compass at points outside the horseshoe / magnet (inside only has been done).	1	Accept 'make measurements outside …'.	8Pe.01 8TWSp.04 AO3
9(a)	Points to include: coil / wind wire around rod connect wire to battery.	1	Both points for one mark.	8Pe.02 AO1
9(b)	Idea that 2 turns are needed to lift 1 paperclip. Including working e.g. 12/6 = 2 Prediction (20/2) = 10 turns	1 1	Must include working or ratio statement. Statement of '10' only for one mark.	8Pe.03 8TWSa.02 8TWSa.03 8TWSp.03 AO2/AO3
9(c)	The strength of an electromagnet is increased by it having an iron core.	1		8Pe.02 8Pe.03 AO2

The Earth in space

Qu	Answer	Mark	Further Information	Spec/TWS/AO
1(a)	(magnetic) compass	1	Accept 'suspended / freely moving bar magnet'.	8ESp.01 AO1
1(b)	nickel, iron	1	Both identified for one mark.	8ESp.01 AO1
2(a)	rice – as food crude oil – refined to make plastics hot rocks (geothermal energy) – to heat homes	2	Two marks for all three correct. One mark for one or two correct.	8ESp.02 AO1
2(b)	rice hot rocks (geothermal energy)	1	One mark for both correct.	8ESp.02 AO1
3(a)	changes	1		8ESc.01 AO1
3(b)	A – glacial period	1		8ESc.01 8TWSa.02 AO2
3(c)	410 000 (years ago)	1	Accept any value between 415 000 and 405 000 years ago.	8ESc.01 8TWSa.05 AO2
3(d)	Animals (tend to) move towards warmer areas when the temperature changes.	1	Accept any appropriate form of words making a similar point.	8ESc.01 AO1
4(a)	asteroid – 1 star – 3 black hole – 4 interstellar dust cloud – 5	2	Two marks for all four correct. One mark for two or 3 correct. No marks for one or none correct.	8ESs.01 8ESs.02 AO2
4(b)	They are too small / not enough mass to hold gases by gravity.	1		8ESs.02 AO1
4(c)	$\dfrac{200 + 250 + 180}{3}$ billion = 210 billion	1 1	One mark for working stated correctly, even if result incorrect. Accept stated result without working for one mark only.	8ESs.01 AO2
5(a)	Carbon dioxide is a greenhouse gas, so it causes the average temperature of the Earth to increase.	1 1		8ESc.02 AO1

5(b)	Any one of: • makes results more reliable • to calculate a mean • idea of identifying anomalous results • idea of convincing others.	1		8ESc.02 8TWSc.03 AO3
6(a)	The weather around the school changes each day.	1	Accept other suitable wording but must include 'weather' and not 'climate.'	8ESc.03 8TWSa.03 AO3
6(b)	Any two from: • measurements taken over long period of time (e.g. years) • measurements taken over much larger area (e.g. country) • averages taken to compare results over long periods of time.	2	One mark per point made.	8ESc.03 8TWSa.04 AO3

End of Year Test 1

Qu	Answer	Mark	Further Information	Spec/TWS/AO
1(a)	$$\text{speed} = \frac{\text{distance}}{\text{time}}$$	1		8Pf.01 8TWSm.03 AO1
1(b)	(400 ÷ 20.0 =) 20	1	Accept value without working or units for 1 mark total. Allow ECF from 1a.	8Pf.01 AO2
	m/s *or* metres per second	1		
2(a)	X: nucleus	1	Allow proton(s)	8Cm.01 AO1
	Y: electron	1		
2(b)	The particles labelled Y are kept in orbit of X because they have opposite **(electric) charge**.	1		8Cm.01, 8Cm.03 AO1
	Most of the space in an atom is **empty**.	1		
3(a)	white blood cells	1		8Bs.02 AO1
3(b)	plasma	1		8Bs.02 AO1
4(a)	helium	1		8ESs.01 AO1
	hydrogen	1		
4(b)	Attracted / pulled to the Earth (as it formed) by gravity.	1		8ESs.01 AO2
5(a)	solubility	1		8Cp.01 8TWSc.01 AO2
5(b)	volume of water	1	Accept amount of water. Accept other sensible answers.	8Cp.01 8TWSp.04 AO3
5(c)	64.5	1		8Cc.05 8TWSp.03 AO2
5(d)	Higher temperature means water particles hit the solid / solute more often.	1	Accept particles have more energy at higher temperature.	8Cc.05 8TWSp.03 AO3

6(a)	carbohydrates – source of energy fats – store of energy proteins – growth and repair water – carrying substances around body and temperature control	2	Two marks for all four correct. One mark for two or three correct. No marks for none or one correct.	8Bp.01 AO1
6(b)	Any one from: • vitamins • minerals.	1	Award one mark if both are stated. Do not accept fibre (it is not a nutrient).	8Bp.01 AO1
6(c)	amount = $20.0 \times 30 \div 100 =$ 6.0 (g)	1 1	Allow answer stated without working for one mark only. Allow 6 or 6.0.	8Bp.01 8TWSa.05 AO2
7(a)	unit of pressure	1		8Pf.05 8TWSc.07 AO3
7(b)	(pressure in) N/m^2	1	Accept $N\ m^{-2}$.	8Pf.05 8TWSc.07 AO1
7(c)	B: 50 C: 25	1 1		8Pf.05 8TWSc.07 AO2
7(d)	Moving particles (in gas) collide with the walls.	1		8Pf.06 AO1
8(a)	oxygen + glucose → water + carbon dioxide	2	One mark per correct substance.	8Bp.05 AO1
8(b)	mitochondrion	1	Accept mitochondria.	8Bp.04 AO1
9(a)	Gas bubbles show the mixture of gases in atmosphere long ago (when they were trapped in the ice).	1		8ESc.01 AO1
9(b)(i)	20 000 years ago	1	Accept ±5 000 years.	8ESc.01 8TWSa.03 AO2
9(b)(ii)	(In colder climate) more water is frozen in sea ice / glaciers, so less is in the sea.	1		8ESc.01 AO2

10(a)	percentage = $\frac{2}{10} \times 100 =$ 20%	1 1	Accept similar but differently arranged working.	8Be.03 8TWSa.05 AO2
10(b)	Any one from these ideas: • outcompete / 'steal' resources from native species • prey on native species.	1		8Be.03 AO1
11(a)		Object	No effect	Attracts	Repels
---	---	---	---
iron filings		✓	
wooden ruler	✓		
steel paperclip		✓	
table tennis ball	✓		| (Effect of lodestone on object)	2	Two marks for all four correct. One mark for two or three correct. No marks for one or zero correct.	8Pe.01 8TWSp.03 AO2
11(b)	iron filings	1		8Pe.01 AO1
12(a)	bones: (wooden) rulers muscles: elastic bands	1	One mark only if both correct.	8Bs.01 8TWSm.01 AO3
12(b)	When elastic band 1 is shortened, elastic band 2 gets longer and ruler B rises. When elastic band 2 is shortened, elastic band 1 gets longer and ruler B moves down.	1 1		8Bs.01 8TWSm.02 AO3

13(a)	green	1		8Ps.04 AO1
13(b)	Points include: • red: one white light and a red filter • green: one white light and a green filter • yellow: one white light with a green filter and a red filter.	2	One mark for explanation of red and green. One mark for explanation of yellow. Deduct one mark if extra equipment is suggested.	8Ps.04 8TWSc.02 AO3
14(a)	C	1	Do not accept any other answer (e.g. if C, D are both listed, award zero marks).	8Cc.02 8TWSa.02 AO3
14(b)	$\left(\dfrac{(-5)+(-3)+(-4)}{3}\right) =$ -4	1 1	Accept other correct expressions (e.g. calculates using positive numbers but correctly adds minus sign in answer). Award one mark if working correct but answer incorrect.	8Cc.02 8TWSa.02 AO3
14(c)	endothermic	1		8Cc.02 8TWSa.03 AO3
14(d)	Any one from: • check starting temperature so it is not unusually high; if too high, allow to cool • use equipment that is at room temperature at start • do not heat the solutions • observe temperature on thermometer more carefully.	1	Accept other appropriate and relevant suggestion.	8Cc.02 8TWSa.04 AO3
14(e)	Any one from: • cools muscle • lowers muscle temperature • coldness reduces pain.	1		8Cc.02 AO2

End of Year Test 2

Qu	Answer	Mark	Further Information	Spec/TWS/AO
1(a)	ball and socket	1		8Bs.01 AO1
1(b)	Allows movement in many different directions.	1		8Bs.01 AO1
2(a)	P	1		8ESs.02 AO2
2(b)	Mars and Jupiter	1	In either order. one mark for both correct. Zero marks for one or none correct.	8ESs.02 AO1
3(a)	Any one from: • shape the wire into a coil • coil the wire around the cylinder.	1		8Pe.02 8TWSc.02 AO3
3(b)	iron	1		8Pe.03 8TWSc.01 AO3
4(a)	tiny plants → small fish	1 1		8Be.01 AO2
4(b)	osprey	1		8Be.02 AO2
5(a)	neutron (particle) positive (charge)	1 1		8Cm.02 AO1
5(b)	Any **one** idea from: • cloud of positive charge is incorrect because positive charges all together in nucleus / at centre • electrons and positive charges mixed evenly is incorrect because electrons orbit positive nucleus • evenly filled space is incorrect because most of an atom is empty space.	1		8Cm.01 8TWSm.02 AO2
6(a)	anticlockwise moment = 6 N m clockwise moment = 6 N m	1 1	One mark per correct answer. Answer must include correct units. If values correct but no units given, allow one mark in total.	8Pf.04 AO2

Qu	Answer	Mark	Further Information	Spec/ TWS/AO
6(b)	balanced	1		8Pf.03 AO1
6(c)	Beam will move clockwise.	1	Accept right-hand end of beam will tip/move down.	8Pf.03 8TWSp.03 AO3
7(a)	wind turbines	1		8ESp.02 AO1
7(b)	100 – 55 – 13 = 32%	1 1		8ESp.02 AO2
8(a)	carbohydrates fats	1	Both nutrient types needed for one mark. If any other nutrient types are listed, award zero marks.	8Bp.02 AO1
8(b)	oxygen + glucose → carbon dioxide + water	1	Both substances needed for one mark.	8Bp.05 AO2
9(a)	Reflected ray drawn at correct angle to normal.	1		8Ps.01 AO2
9(b)(i)	(Image) same size / shape as candle.	1		8Ps.01 AO2
9(b)(ii)	Appears to be 50 cm behind mirror.	1		8Ps.01 AO2
9(c)	D	1		8Ps.02 8TWSp.03 AO2

Qu	Answer	Mark	Further Information	Spec/ TWS/AO
9(d)	C	1		8Ps.02 8TWSp.03 AO2
10(a)	Table **X** only.	1		8ESc.03 8TWSa.03 AO1
10(b)	$\dfrac{23+25+30+26}{4} =$ 26 °C	1 1		8ESc.03 AO2
10(c)	Average temperature is rising steadily / slowly.	1		8ESc.02 8TWSa.02 AO2
11(a)	Any two from: • wear eye protection • wear protective clothing / lab coat • use tongs / test tube holder • point open end of test tube away from yourself and other people • place hot test tube in rack to allow it to cool • turn Bunsen burner off when experiment is complete.	2	One mark per correct point. Accept any other relevant safety precaution.	8Cc.03 8TWSp.05 AO3
11(b)	M > N > L	1		8Cc.03 8TWSa.02 AO3
11(c)	Yes (L is silver) because L is inert / does not react with oxygen even when heated.	1 1	Do not accept 'very shiny' as a reason (this is supporting evidence, but is not sufficient on its own).	8Cc.03, 8Cc.06 8TWSa.01 AO3
11(d)	calcium + oxygen → **calcium oxide**	1		8Cc.01 AO2
12(a)	Variable that is kept the same during an investigation.	1		8TWSp.04 AO3
12(b)	Any one from: • same clothes throughout run • temperature	1	Accept any sensible suggestion.	8TWSp.04 AO3

Qu	Answer	Mark	Further Information	Spec/TWS/AO
	• height of ground (level throughout).			
12(c)	Straight line that passes through, and does not run beyond the points (0,0) and (60 minutes, 12 km).	1 1		8Pf.02 8TWSa.05 AO3
13(a)	Any **one** from: • a pure substance would produce only one spot on the chromatogram • two spots on the chromatogram show that two substances are present.	1		8Cp.02 8TWSa.03 AO1
13(b)	copper chloride	1		8Cp.02 8TWSc.01 AO3
13(c)	Measure the melting / boiling point of the copper sulfate. If an impurity is present, it will change the melting / boiling point (compared to pure copper sulfate).	1 1		8Cc.04 8TWSa.04 AO3

Qu	Answer	Mark	Further Information	Spec/TWS/AO
14(a)	breathing	1		8Bs.03 AO1
14(b)(i)	exhaling	1		8Bs.03 AO1
14(b)(ii)	Diaphragm rises. Arrow linked to diaphragm pointing upwards.	1	Both correct for one mark.	8Bs.03 AO1
14(c)(i)	Any **two** of: • smoking reduced lung capacity • the people who have smoked for over 10 years have much lower lung capacity than the people who have never smoked • stopping smoking helps restore lung capacity, but it takes many years.	2	One mark for each point up to a maximum of two.	8Bp.03 8TWSa.03 AO3
14(c)(ii)	Any **one** from: • data may be biased compare results with other information to avoid bias.	1	Also accept comments about reliability of results and/or improving reliability.	8Bp.03 8TWSc.06 AO3

Periodic Table

1	2												3	4	5	6	7	0
							1 **H** hydrogen 1											4 **He** helium 2
7 **Li** lithium 3	9 **Be** beryllium 4												11 **B** boron 5	12 **C** carbon 6	14 **N** nitrogen 7	16 **O** oxygen 8	19 **F** fluorine 9	20 **Ne** neon 10
23 **Na** sodium 11	24 **Mg** magnesium 12												27 **Al** aluminium 13	28 **Si** silicon 14	31 **P** phosphorus 15	32 **S** sulfur 16	35.5 **Cl** chlorine 17	40 **Ar** argon 18
39 **K** potassium 19	40 **Ca** calcium 20	45 **Sc** scandium 21	48 **Ti** titanium 22	51 **V** vanadium 23	52 **Cr** chromium 24	55 **Mn** manganese 25	56 **Fe** iron 26	59 **Co** cobalt 27	59 **Ni** nickel 28	63.5 **Cu** copper 29	65 **Zn** zinc 30		70 **Ga** gallium 31	73 **Ge** germanium 32	75 **As** arsenic 33	79 **Se** selenium 34	80 **Br** bromine 35	84 **Kr** krypton 36
85 **Rb** rubidium 37	88 **Sr** strontium 38	89 **Y** yttrium 39	91 **Zr** zirconium 40	93 **Nb** niobium 41	96 **Mo** molybdenum 42	[98] **Tc** technetium 43	101 **Ru** ruthenium 44	103 **Rh** rhodium 45	106 **Pd** palladium 46	108 **Ag** silver 47	112 **Cd** cadmium 48		115 **In** indium 49	119 **Sn** tin 50	122 **Sb** antimony 51	128 **Te** tellurium 52	127 **I** iodine 53	131 **Xe** xenon 54
133 **Cs** caesium 55	137 **Ba** barium 56	139 **La*** lanthanum 57	178 **Hf** hafnium 72	181 **Ta** tantalum 73	184 **W** tungsten 74	186 **Re** rhenium 75	190 **Os** osmium 76	192 **Ir** iridium 77	195 **Pt** platinum 78	197 **Au** gold 79	201 **Hg** mercury 80		204 **Tl** thallium 81	207 **Pb** lead 82	209 **Bi** bismuth 83	**Po** polonium 84	**At** astatine 85	**Rn** radon 86
Fr francium 87	**Ra** radium 88	**Ac**** actinium 89	**Rf** rutherfordium 104	**Db** dubnium 105	**Sg** seaborgium 106	**Bh** bohrium 107	**Hs** hassium 108	**Mt** meitnerium 109	**Ds** darmstadtium 110	**Rg** roentgenium 111								

Key:
relative atomic mass
atomic symbol
name
atomic (proton) number

Elements with atomic numbers 112–116 have been reported but not fully authenticated

La lanthanoids

Ac actinoids

Glossary

Biology

absorb: to take in or soak up.
addictive: substance that makes people feel that they must have it.
adolescence: the life stage in humans that usually happens between the ages of 11 and 18. During this stage, people go through many emotional and physical changes.
aerobic respiration: respiration that requires oxygen to release energy from glucose.
alimentary canal: tube that runs from your mouth to your anus.
alveolus: tiny, pocket-shaped structure in lungs where gaseous exchange happens. The plural is alveoli.
amniotic fluid: liquid that surrounds the foetus in the uterus.
anus: last organ in the alimentary canal. Faeces leave your body here.
aorta: large artery that leaves the left ventricle of your heart.
artery: thick-walled blood vessel that carries blood away from the heart.
atria: chambers at the top of your heart. You have a left atrium and a right atrium.
balanced diet: eating many different foods to get the correct amounts of nutrients.
bile: liquid that helps fat-digesting enzymes to work.
biomass: mass of all the compounds in an organism, which it has made.
blood: liquid organ that carries substances around the body.
blood vessels: tube-shaped organs that carry blood around the body.
breathing: movements of muscles in your respiratory system that cause air to move in and out of your lungs.
breathing rate: the number of times you inhale and exhale in one minute.
bronchioles: small tubes leading from the bronchus in a lung.
bronchus: large tube leading from the trachea into a lung. Plural is bronchi.
cancer: when cells in a tissue start to make many copies of themselves very quickly.

capillary: tiny blood vessel that carries blood from arteries to veins.
carbohydrate: compound made from carbon, hydrogen and oxygen.
catalyst: substance that speeds up a chemical reaction.
chamber: space inside the heart that fills with blood and empties again.
chemical digestion: digestion that is done by chemical substances.
chemical reaction: change in which new substances are produced.
chest: area inside the body between the ribcage, neck, backbone and diaphragm.
chlorophyll: green substance that absorbs light, to get energy for photosynthesis.
chloroplast: green part of a plant cell that contains chlorophyll.
cilia: waving strands that stick out of some cells.
ciliated epithelial cell: specialised cell with waving cilia to sweep mucus along.
circulation: movement of blood around the body.
circulatory system: group of organs that gets blood around the body.
clot: thick mass of blood cells, stuck together.
compound: substance made from elements.
constipation: when your intestines become blocked.
contract (muscle): when muscle tissue gets shorter and fatter, it contracts.
control variable: variable that you keep the same during an investigation.
correlation: relationship (link) between variables where one increases or decreases as the other increases.
cuticle: waterproof covering on leaves.
deficiency disease: another term for nutritional deficiency.
diaphragm: organ that helps breathing.
diet: what you normally eat or drink.
diffusion: the spreading out of particles from where there are many (high concentration) to where there are fewer (lower concentration).

digestive juice: liquid that contains enzymes to digest food.
double circulatory system: circulatory system in which the heart pumps blood around two circuits. In humans, one circuit supplies the lungs, and the other supplies the rest of the body.
drug: substance that affects the way your body works.
element: substance that contains only one type of atom; it cannot be split into anything simpler.
embryo: small ball of cells that develops from a fertilised egg cell. It becomes attached to the uterus lining and develops into a foetus.
enzyme: substance that digests food.
epidermis cell: cell that forms an outer covering of a leaf, to protect the leaf.
excrete: getting rid of wastes made inside an organism.
exhale: breathing out.
faeces: solid waste material produced by humans and other animals.
fats: nutrients needed by your body to store energy.
fibre: food substance that cannot be digested but which helps to keep your intestines healthy.
foetus: baby developing in a woman's uterus, from about 10 weeks of development, when it starts to resemble a baby.
gall bladder: organ next to your liver that stores bile.
gaseous exchange: when two or more gases move from place to place in opposite directions.
gestation period: the amount of time for a baby to fully develop in the mother's uterus. This time varies for different animals – for example, 31 days for rabbits and 22 months for elephants.
glucose: sugar made by digesting carbohydrates (in animals) and by photosynthesis (in plants).
growth hormone: chemical made in the brain that causes growth in the body.
guard cell: cell that helps form a stoma in a leaf, to allow gases in and out.
gullet: another word for oesophagus.

gut: another word for alimentary canal.
haemoglobin: substance that collects oxygen.
hazard: harm that something may cause.
heart: organ that pumps blood through blood vessels.
heart attack: when heart muscle cells start to die and the heart does not pump properly.
heartbeat: squeezing of the muscles in the heart wall to push blood into blood vessels.
heart rate: the number of heartbeats in one minute.
high blood pressure: when blood puts too much pressure on blood vessels.
hormone: chemical released into the bloodstream, which has an effect on certain parts of your body.
implantation: the developing embryo attaches to the uterus lining.
infant: the life stage which lasts for the first year after birth.
inhale: breathing in.
iodine solution: liquid that turns from orange to blue-black when added to starch.
joule: unit used to measure energy.
kwashiorkor: deficiency disease caused by a lack of protein.
large intestine: organ of the alimentary canal. It removes water from undigested food to make faeces.
line of best fit: straight or curved line drawn through the middle of a set of points to show the pattern of data points.
lipids: another word for fats.
liver: organ that makes and destroys substances. It makes bile.
lungs: organs that get oxygen into the blood and remove carbon dioxide.
malnutrition: when a diet contains too much or too little of something, and causes health problems.
mechanical digestion: digestion that is done by physical actions, such as chewing.
menopause: the time in a woman's life when the menstrual cycle stops, normally between ages 45 and 55.
menstrual cycle: cycle of changes that happens in females after puberty. During each cycle an egg cell is released from an ovary and (if it is not fertilised) menstruation happens. Each cycle lasts between about 24 and 35 days.

menstruation: the time in the menstrual cycle when the uterus lining breaks down and is lost. It is called a period.
microscopic: something so small you can only see it using a microscope.
mineral salt: type of substance in the soil that plants need small amounts of. Often just called a 'mineral'.
minerals: nutrients that living organisms need in small amounts for health, growth and repair. Also called mineral salts.
mitochondria: part of a cell where aerobic respiration happens to release energy. The singular form is 'mitochondrion'.
model: simple way of showing or explaining a complicated object or idea.
molecule: group of two or more atoms joined together. Oxygen, carbon dioxide and water all exist as molecules.
mouth: first organ in the alimentary canal.
mucus: sticky liquid that traps particles.
nutrient: substance you need in your diet for energy or as a raw material.
nutritional deficiency: problem caused by a lack of a nutrient in the diet. Also called a deficiency disease.
obesity: being so overweight that your health is in danger.
oesophagus: organ of the alimentary canal. Its muscle walls push food from your mouth into your stomach.
oestrogen: hormone that triggers many of the physical changes in girls during puberty.
ovary: female reproductive organ where eggs are made, stored and matured. Females have two ovaries.
oviduct: the tube which connects the ovary to the uterus. Females have two oviducts, one from each ovary.
ovulation: the release of an egg cell from one of the ovaries.
palisade cell: cell that contains many chloroplasts for photosynthesis.
pancreas: organ that makes enzymes to digest fats, proteins and carbohydrates.
period: stage in the menstrual cycle when the lining of the uterus is lost from the body.
peristalsis: contraction and relaxation of muscles in the alimentary canal that pushes food along.

pharmaceutical drug: drug used in healthcare to help the body fight a disease, or to relieve pain.
phloem cell: plant cell that is adapted to form living tubes to transport sugars and other substances.
photosynthesis: chemical reaction that plants use to make their own food.
placenta: organ which forms in the uterus, linking the developing foetus to the uterus wall (and therefore the mother).
plaque: lump of fatty material that builds up inside an artery.
plant vein: tube containing smaller tubes that carry substances around a plant.
plasma: liquid part of the blood.
platelet: cell fragment that helps your blood to clot.
pregnant: a woman becomes pregnant if a fertilised egg implants in her uterus.
product: substance made during a chemical reaction.
prostate gland: gland that surrounds the bottom of a male's bladder. It produces some of the liquid that makes up semen.
proteins: nutrients you need for growth and repair.
puberty: the physical changes that happen to the body during adolescence.
pulse: wave of stretching along the wall of an artery each time the heart beats.
raw material: another term for reactant.
reactant: substance that changes in a chemical reaction to form products.
rectum: organ of the alimentary canal. It stores faeces.
red blood cell: cell that contains haemoglobin so it can carry oxygen.
relax (muscle): when muscle tissue stops contracting, it relaxes.
respiration: chemical process that happens in all parts of an organism to release energy.
respiratory system: group of organs that get oxygen into the blood and remove carbon dioxide.
rib: bone that helps to protect your heart and lungs.
ribcage: all your ribs.
rickets: deficiency disease caused by a lack of calcium or vitamin D.
risk: chance of a hazard causing harm.

root hair cell: plant cell found in roots that is adapted for taking in water quickly.
saliva: digestive juice made by salivary glands in your mouth.
salivary gland: organ inside your mouth that makes saliva.
scatter graph: graph of two variables, both measured in numbers.
scurvy: deficiency disease caused by a lack of vitamin C.
small intestine: organ of the alimentary canal. It makes enzymes and lets digested food pass into your blood.
spongy cell: irregularly shaped cell that helps form air spaces in a leaf.
starch: large carbohydrate, which plants use to store energy.
stillborn: the term used to describe a baby that is dead when it is born.
stoma: hole in a leaf, formed between two guard cells. The plural is stomata.
stomach: organ of the alimentary canal. It makes enzymes to digest proteins and churns food into a smooth soup.
stroke: when brain cells die due to a lack of blood (which is caused by a blocked blood vessel in the brain).
sugar: type of small carbohydrate.
surface area: the area of a surface, measured in squared units such as square centimetres (cm^2).
symptom: effect of a disease on the body.
testosterone: hormone that triggers many of the physical changes in boys during puberty.
tissue: group of cells of the same type.
trachea: tube-shaped organ that allows air to flow in and out of your lungs.
transpiration: the flow of water into a plant's root, up its stem and out of its leaves.
trend: pattern seen in data.
tumour: a lump of cancer cells.
type 2 diabetes: disease that may damage organs.
umbilical cord: flexible tube containing blood vessels from the foetus – it connects the foetus to the placenta.
urea: waste product made by the liver and excreted by the kidneys.
uterus: female reproductive organ where the baby grows when a woman is pregnant.
valve: flaps of tissue that only allow blood to flow in one direction.
variable: something that may change in an experiment.
vein: thin-walled blood vessel that carries blood towards the heart.
ventricle: chamber at the bottom of your heart. You have a left ventricle and a right ventricle.
vitamins: nutrients you need in small amounts for health, growth and repair.
white blood cell: cell that helps destroy microorganisms.
wilting: when a plant becomes floppy due to lack of water.
word equation: model showing what happens in a chemical reaction, with reactants on the left of an arrow and products on the right.
xylem cell: plant cell that is adapted to form hollow tubes to transport water.
zygote: fertilised egg cell.

Chemistry

alloy: mixture of metal with other elements.
atom: the smallest particle of a substance (element) that can exist and still be the same substance.
carbonate: compound that reacts with an acid to give carbon dioxide, a salt and water. For example, calcium carbonate ($CaCO_3$).
chemical reaction: a change in which new substances are produced.
chemical symbol: short way of representing an element's name.
chloride: salt that is formed when hydrochloric acid reacts with another element; for example, sodium chloride (NaCl).
combustion: chemical reaction between a substance and oxygen, which transfers energy as heat and light.
compound: contains atoms or more than one element strongly held together. Compounds have different properties to the elements they contain.
concentration: measure of how many particles of a certain type there are in a volume of liquid or gas.

correlation: relationship (link) between variables where one increases or decreases as the other increases.
corrosion: the damage of metals through chemical reactions with substances in the air or water.
dense: has a high mass in a small volume.
diffusion: the spreading out of particles from where there are many (high concentration) to where there are fewer (lower concentration).
distillation: separation method used to separate a liquid from a mixture.
ductile: able to be stretched into wires.
element: substance that contains only one type of atom; it cannot be split into anything simpler.
evaporation: method used to separate a soluble solid from a liquid.
filtration: method used to separate an insoluble solid from a liquid.
formula: shows the chemical symbols of elements in a compound, and how many of each type of atom there are.
gas pressure: the effect of the forces caused by collisions from gas particles on the walls of a container.
hazard: harm that something may cause.
hydroxide: compound that contains one atom each of oxygen and hydrogen bonded together; for example, potassium hydroxide (KOH).
insoluble: substance that does not dissolve.
magnetic: material that is attracted by a magnet.
mixture: two or more elements or compounds mixed together. They can easily be separated.
molecule: group of two or more atoms joined together. Oxygen, carbon dioxide and water all exist as molecules.
oxidation: chemical reaction with oxygen to form a compound that contains oxygen.
oxide: compound that is formed when oxygen reacts with another element; for example, magnesium oxide (MgO).
Periodic Table: list of all the elements.
product: substance made during a chemical reaction.
pure: substance that contains only one element or compound.
rate: measurement of how quickly something happens.
reactant: substance that changes in a chemical reaction to form products.
reactivity: how likely it is that a substance will undergo a chemical reaction.
risk: chance of a hazard causing harm.
rusting: chemical reaction of iron with oxygen and water. It is the corrosion of iron.
salt: compound formed when an acid reacts with a base or a metal.
soluble: substance that dissolves to form a solution.
solution: mixture of a soluble substance and a liquid.
sulfate: salt that is formed when sulfuric acid reacts with another element; for example, copper sulfate ($CuSO_4$).
theory: idea or set of ideas that explains an observation.
word equation: model showing what happens in a chemical reaction, with reactants on the left of an arrow and products on the right.

Physics

absorption: the way in which an object takes in the energy reaching its surface.
accurate: accurate measurement is where the measurement is very close to the real value.
amplitude: the maximum height of the wave, from the centre to the top or bottom.
angle of incidence: this is the angle between the incident ray and the normal.
angle of reflection: this is the angle between the reflected ray and the normal.
anomalous result: result that does not follow the same pattern as other measurements.
apparent depth: how deep something appears to be.
attract: pull closer together.
average: the mean average of a set of numbers is found by:
total of all the numbers added together divided by how many different numbers there are.
compression: region where particles in a longitudinal wave are closer together.

control variable: the variable that you keep the same during an investigation.
core: piece of metal (usually iron) that a coil of wire is wound around to increase the strength of the magnetic field.
correlation: relationship (link) between variables where one increases or decreases as the other increases.
dependent variable: the variable in an investigation that you measure.
diffraction grating: transparent piece of glass or plastic which has many lines drawn onto it. Light can pass through the spaces between the lines.
dispersion: the splitting of white light into a spectrum of colours.
electromagnet: magnet that can be switched on or off using an electric current.
filter: colour filter will only allow light of its own colour to pass through it.
frequency: the number of waves per second.
gradient: the gradient of a graph tells you how steep the line is.
Hertz: the unit of frequency. 1 Hz = 1 complete wave every second.
incident ray: this ray shows the light travelling towards the mirror.
independent variable: the variable in an investigation that you change.
infrasound: sound waves with a frequency too low for humans to hear.
light ray: straight line which shows the direction of light.
line of best fit: straight or curved line drawn through the middle of a set of points to show the pattern of data points.
longitudinal wave: in a longitudinal wave, the wave travels in the same direction as the vibrations that produced it.
magnetic field: the region around a magnetic material in which a magnetic force acts.
magnetic force: force that occurs when a magnet attracts another object or repels another magnet.
magnetic pole: point on a magnet where the force is strongest.

magnetism: property of some materials that gives rise to forces between these materials and magnets.
medium: the substance a sound wave travels through.
normal: this is a line drawn at 90° to the mirror at the point where rays hit the mirror.
opaque: material that prevents light travelling through.
permanent magnet: object made from a magnetic material that retains its magnetism for a very long time.
pitch: the pitch of a sound is how high or low it is.
plane mirror: plane means flat so a plane mirror is a flat mirror.
primary colours: red, blue and green. Mixing these colours of light together will make all other colours of light.
prism: transparent object that refracts light.
rarefaction: region where particles in a longitudinal wave are further apart.
real depth: how deep something really is.
reflected ray: this shows the light travelling away from the mirror after it has been reflected.
refraction: the bending of light when it enters a different medium.
repel: push further apart.
scattering: scattering happens when light is reflected from particles and uneven surfaces.
secondary colours: yellow, magenta and cyan.
shadow: dark area caused when light is blocked.
solar eclipse: when the Sun looks like it is completely or partly covered with a dark circle.
speed: how far something moves in a given time.
transparent: material that lets light through.
ultrasound: sound waves with a frequency too high for humans to hear.
vacuum: completely empty space.
vibration: when something moves back and forwards many times, we say it vibrates.
wavelength: the length of one complete wave.